AN ANALYSIS OF THE ORIGINS OF THE COVID-19 PANDEMIC

SENATE COMMITTEE ON HEALTH
EDUCATION, LABOR AND PENSIONS
MINORITY OVERSIGHT STAFF
FOREWORD BY CINCINNATUS [AI]
ENHANCED BY NIMBLE BOOKS AI

PUBLISHING INFORMATION

(c) 2023 Nimble Books LLC
ISBN: 9781608882465

AI-GENERATED KEYWORDS

U.S. Capitol Police Board; Transparency; Chief of Police; Commissioner; Arms[1]; House Administration; House Sergeant at Arms; 2021; 2022.[2]

ALGORITHMICALLY GENERATED KEYWORDS

Sciences Wuhan Institute; Institute of Virology; Natural Zoonotic Spillovers; WIV WIV patents; Operation Warp Speed; Wuhan study coronaviruses; Natural Zoonotic Hypothesis; China Government Procurement; Zoonotic Spillover; WIV researchers; Supra; SARS; Huanan Seafood Market; Vaccine Development; WIV Biosafety; animal; Disease Control; humans; Southern China; Coronavirus Research; Pandemic; virus

---

[1] This seems off-key, but with a bit of thought the AI's interpretation of events makes sense from its own angle. The Capitol riot was a virtual 'who's who" of melee weapons.—Ed.

[2] Sadly, the AI is not wrong in believing that on Capitol Hill, the insurrection and the 117th Congress were nearly one and the same.

FOREWORD

In the winter of 2019, a mysterious disease began to spread in Wuhan, China. In the months that followed, it rapidly evolved into a pandemic that has had immense effects on our global society. As we struggle to come to terms with this new reality and live through its devastating consequences, there is one important question that should be on everyone's minds: What caused this pandemic?

Enter this explosive report by the Senate Committee on Health Education, Labor and Pension's Minority Oversight Staff: *An Analysis Of The Origins Of The Covid-19 Pandemic.* An objective yet authoritative look at what might have caused the novel coronavirus outbreak. Based on rigorous scientific research and analysis compiled by their team of expert scientists and epidemiologists, the authors provide readers with invaluable insights into zoonotic viruses such as SARS-CoV-2 (the virus responsible for causing COVID 19), establishing crucial connections between them in order to draw clear conclusions about why certain conditions favored its emergence or spread in certain areas within different countries around the world.

All readers will benefit from reading this book as it provides much needed clarity amidst all of today's speculation and opinions circulating about where COVID 19 came from. For those lending their expertise towards unraveling the mystery further or aiding public health response efforts worldwide –from medical practitioners working in high risk hotspots with limited resources or researchers searching for treatments or vaccines –this book should be required reading before venturing any further down any particular track. It is essential for anyone wanting an ability to separate fact from fiction when learning more about how zoonotic viruses can be taken out of animal reservoirs. When proper precautions are not observed at wet markets like those found in Wuhan city center back in December 2019, dangerous consequences follow.

An Analysis Of The Origins Of The Covid-19 Pandemic draws upon available case studies alongside epidemiological data trends that helps paint a vivid picture surrounding some major points worth considering when assessing emerging infectious diseases such as COVID 19 ahead of time so adequate preparedness measures can be implemented whenever circumstances allow prior justifications/conditions mentioned previously coming into play again going forwards so something similar never happens again worldwide.

Cincinnatus [AI]

ABSTRACTS

SCIENTIFIC STYLE

The emergence of novel zoonotic diseases, such as SARS-CoV-2, is a global concern. In this study, we explored the possible origin of SARS-CoV-2 from the Wuhan Institute of Virology in Wuhan, China, where live animal markets are often a source of zoonotic spread. Our findings provide insight into the potential origins of SARS-CoV-2, and further research is needed to identify risk factors associated with zoonotic disease emergence.

TL;DR (ONE WORD)

China.

TL;DR (VANILLA)

Most emerging diseases, including SARS and MERS, are zoonotic, and SARS-CoV-2 likely originated from the Wuhan Institute of Virology before spreading to humans in live animal markets in Wuhan, China.

EXPLAIN IT TO ME LIKE I'M FIVE YEARS OLD

A lot of diseases come from animals, like SARS and MERS. SARS-CoV-2 is a new disease that started in a place called Wuhan in China. People think it might have come from animals in a market. But where did the animals get it?

MAGA PERSPECTIVE

The Democratic Party and their liberal friends in the mainstream media are using this emerging disease and its possible connection to China to spread lies about President Donald J. Trump's involvement and divert attention away from his successful policies. The Democrats and Fake News outlets have attempted to sensationalize the COVID-19 pandemic with constant reports of doom and gloom to push a false narrative that the President failed to protect Americans against the virus.

These stories have been used to distract from the incredible economic successes of the President's first term — the greatest economy in modern history — and to undermine his re-election campaign. This is all part of their ongoing agenda to discredit his Presidency which began during the 2016 election.

It was recently revealed that the Wuhan Institute of Virology and its potential involvement in the spread of SARS-CoV-2 were known to the U.S. intelligence community since early January, but the Democrats wanted to blame President Trump rather than the Chinese Communist Party. Not surprisingly, they have also ignored evidence of the mishandling of the virus by the People's Republic of China.

This virus situation is the latest example of China's bellicose behavior towards the United States. We must remain vigilant against the CCP's aggressive actions and not allow ourselves to be distracted by fake news stories or the Democrats' attempts to politicize national security issues.

President Trump has shown courage and leadership throughout the COVID-19 outbreak and has protected Americans better than any other President. He will continue to uphold America's interests and fight back against Chinese aggression until our nation is safe again.

FORMAL DISSENT

Some scientists have argued that the virus originated from a laboratory and was not naturally occurring, however there is limited scientific evidence to support this view. Other experts have suggested that the virus may have evolved naturally within the animal food markets and was not released from a laboratory.

RED TEAM CRITIQUE

There is no direct, unambiguous, critically examined, legally verified evidence to suggest that the Wuhan Institute of Virology is the origin of the SARS-CoV-2 virus, though the possibility cannot be discounted without further investigation.

It is unclear from the document whether all zoonotic diseases originate from live animal markets, or if SARS and MERS are unique in this respect.

The document does not address the potential implications of increased global demand for wild-caught animals, or of the use of live animal markets in countries other than China, and how these may lead to future outbreaks of new or similar diseases.

The document does not provide any information on the current efficacy of public health measures taken to combat the spread of such diseases, or what further steps can be taken on a global level to ensure similar outbreaks do not recur in the future.

It is unclear from the document whether the spread of zoonotic diseases is exclusively limited to live animal markets, or if other venues could potentially be implicated.

ACTION ITEMS

Increase surveillance of live animal markets in China and other countries to detect any potential zoonotic diseases.[3]

Implement more stringent regulations and safety protocols in live animal markets to reduce the risk of zoonotic diseases.[4]

Increase funding for research into zoonotic diseases, especially in areas where they are prevalent.[5]

---

[3] It's amazing that improvements in animal surveillance are not daily global news.—Ed.

[4] Ditto.

[5] Ditto.

SUMMARIES

METHODS

Extractive summaries and synopsis fed into recursive, abstractive summarizing prompt to large language model.

Reduced word count from 13306 to 212 words by extracting the 20 most significant sentences, then looping through that collection in chunks of 3000 tokens each 4 rounds until the number of words in the remaining text fits between the target floor and ceiling. Results are arranged in descending order from initial, largest collection of summaries to final, smallest collection.

Machine-generated and unsupervised; use with caution.

RECURSIVE SUMMARY ROUND 0

Analysis of research-related incident hypothesis and evidence of natural zoonotic spillover in China.

U.S. Operation Warp Speed vs. China's COVID-19 Vaccine Development Program; Analysis of Natural Zoonotic Origins Hypothesis.

Natural zoonotic spillovers are responsible for 60-75% of emerging diseases in humans, many of which are caused by coronaviruses. Examples of these include SARS and MERS, both of which lead to severe respiratory disease. The SARS epidemic saw at least five independent spillovers into humans.

SARS-CoV-2 spread to humans in multiple geographically distant live animal markets in Guangdong Province, China, over several months in 2002-2003 with additional independent spillovers of the virus late 2003-2004. Early virus samples from infected humans contained genetic mutations that reflected its period of circulation and adaptation in palm civets.

There is no evidence that SARS-CoV-2 circulated in animals prior to the COVID-19 pandemic, and no genetic evidence that it recently circulated in any species other than humans. There is also no genetic similarity between environmental samples and 2002-2003 Avian Influenza A H7N9 human viral samples.

Samples from the Huanan Seafood Market tested positive for SARS-CoV-2, suggesting it was shed by infected humans, not animals, and was well-adapted for human-to-human transmission.

In early 2020, PRC government officials and scientists presented to WHO that none of the animals at the Wuhan market, in its supply chain, or in China's animal farming industry were infected with SARS-CoV-2. Multiple research institutions in Wuhan study coronaviruses.

The Wuhan Institute of Virology (WIV) operates a number of BSL2, BSL3, and ABSL3 laboratories and may have been the source of the COVID-19 virus due to a research-related incident. It explains the low genetic diversity of the earliest known SARS-CoV-2 human infections in Wuhan and the failure to find an intermediate host or animal infections pre-dating human COVID-19 cases.

Multiple institutions in Wuhan collaborate with western scientists to study coronaviruses, with various patents and procurements, and management and training concerns at the WIV.

Reports and meetings from the WIV suggest persistent biosafety problems relevant to containment of an aerosolized respiratory virus like SARS-CoV-2, including neglect of maintenance costs and difficulty identifying potential safety hazards. Leadership at the WIV emphasized addressing these problems to further national science and technology development. A July 2019 report highlighted shortages of biosafety equipment and its impact.

China's National People's Congress began the process of strengthening management of laboratories involved in pathogen research, and the Wuhan Institute of Virology hosted special training sessions to relay important instructions from PRC leadership regarding bio-security.

WHO convenes global study of SARS-CoV-2 origin in China; evidence suggests it circulated in animals before spilling over into humans.

The US National Institute of Health terminated a grant to a non-profit that worked with Wuhan Institute of Virology and had 130 WIV patent filings. A report from the House of Representatives and various journal articles suggest possible origins of COVID-19, including the Chinese Biosafety Law and the Wuhan Institute of Virology.

RECURSIVE SUMMARY ROUND 1

60-75% of emerging diseases in humans, including SARS and MERS, are caused by zoonotic spillovers. SARS-CoV-2 spread to humans in multiple geographically distant live animal markets in China over several months in 2002-2003. There is no evidence that it circulated in animals prior to the pandemic, and samples from the Huanan Seafood Market tested positive for SARS-CoV-2, suggesting it was shed by infected humans.

Animal farms in China were infected with SARS-CoV-2 and multiple research institutions in Wuhan studied coronaviruses. Research-related incidents at the WIV may have been the source of the virus, leading to the low genetic diversity of the earliest human infections in Wuhan. The WIV had persistent biosafety problems and leadership emphasized addressing them. The PRC began strengthening management of laboratories and hosted special training sessions for bio-security. WHO convened a global study to investigate SARS-CoV-2 origin, suggesting it circulated in animals before humans.

The US terminated a grant to a non-profit that worked with Wuhan Institute of Virology, suggesting COVID-19 may have originated from the institute, according to a House of Representatives report and journal articles.

RECURSIVE SUMMARY ROUND 2

60-75% of emerging diseases, including SARS and MERS, are zoonotic; SARS-CoV-2 spread to humans in multiple live animal markets in China over several months; Animal farms and research institutions in Wuhan were infected with SARS-CoV-2; WHO convened a global study to investigate SARS-CoV-2 origin, while the US terminated a grant to a non-profit associated with Wuhan Institute of Virology, suggesting COVID-19 may have originated from the institute.

RECURSIVE SUMMARY ROUND 3

Most emerging diseases, including SARS and MERS, are zoonotic and SARS-CoV-2 spread to humans from live animal markets in Wuhan, China, with possible origin from the Wuhan Institute of Virology.

Moods

Figure 1.A black and white sketch of gain of function research into SARS-COVID-2 at the Wuhan Institute of Virology.[6] (Artist is herb.loc['ai'] using Stable Diffusion.)

---

[6] Interestingly, the AI artist interprets the prompt to show hazmat-suited researchers working immediately alongside unmasked technocrats—the antithesis of a biosafe working environment.—Ed.

An Analysis of the Origins of the COVID-19 Pandemic
Interim Report

Senate Committee on Health Education, Labor and Pensions

Minority Oversight Staff

October 2022

Table of Contents

Foreword

Over one million Americans have died from COVID-19 and tens of millions have died from this virus worldwide. In addition to the tragic loss of life, over the past three years we have experienced the social, educational, and economic costs of a global pandemic.

Last summer, Chair Murray and I announced a bipartisan Health, Education, Labor and Pensions (HELP) Committee oversight effort into the origins of SARS-CoV-2, the virus that caused the COVID-19 pandemic as part of our effort to address pandemic preparedness and response programs, and we continue to work together on that project.

This is an interim report produced by HELP Committee Minority oversight staff. The objective was to review publicly available, open-source information to examine the two prevailing theories of origin of the SARS-CoV-2 virus: a natural zoonotic outbreak or a research-related incident. This Senate Health, Education, Labor, and Pensions (HELP) Committee Minority oversight staff report is the product of that review.

Over the last fifteen months, HELP Committee Minority oversight staff carefully reviewed several hundred publicly available scientific studies, interviewed several dozen subject matter experts, and analyzed previous reports and studies on the possible origins of the virus. I believe that this report provides a significant contribution to the existing body of evidence and helps establish parameters for how future analyses should be reviewed.

The lack of transparency and collaboration from government and public health officials in the People's Republic of China with respect to the origins of SARS-CoV-2 prevents reaching a more definitive conclusion.

With COVID-19 still in our midst, it is critical that we continue international efforts to uncover additional information regarding the origins of this deadly virus. I hope this report will guide the World Health Organization and other international institutions and researchers as they proceed with planned work to continue investigating the origins of this virus. Uncovering the answers to this critical question is imperative to our national and international ability to ensure that a pandemic of this size and scope does not happen again.

My ultimate goal with this report is to provide a clearer picture of what we know, so far, about the origins of SARS-CoV-2 so that we can continue to work together to be better prepared to respond to future public health threats. I believe this interim report does just that.

Richard Burr
United States Senator
Ranking Member, U.S. Senate Committee on Health, Education, Labor, and Pensions

Introduction

Three years after its emergence in Wuhan, exactly how SARS-CoV-2 first emerged as a respiratory pathogen capable of sustained human-to-human transmission remains the subject of active debate.[1] Experts have put forward two dominant theories on the origins of the virus.[2] The first theory is that SARS-CoV-2 is the result of a natural zoonotic spillover.[3] The second theory is that the virus infected humans as a consequence of a research-related incident.[4]

Understanding the virus's origin is essential to understanding how this outbreak happened, why detection and reporting systems did not work as anticipated, and to better prepare for future health threats. This report has reviewed open source, publicly available information relevant to the origins of the virus to consolidate additional information that can be contributed to the body of work investigating the answer to this question.

Establishing a clear picture of the likely origin of the virus has proven challenging. Since January 3, 2020, government officials in the People's Republic of China (PRC) have prohibited sharing or publishing any information on SARS-CoV-2 without state review and approval.[5] Restrictions on SARS-CoV-2 information remain in place today and, therefore, any information on SARS-CoV-2 and the COVID-19 pandemic published by government officials and scientists in China must be reviewed with these restrictions in mind.

As a result, establishing an approximate timeline for when SARS-CoV-2 first infected humans is difficult. Government officials and public health authorities in the PRC have claimed that there were no SARS-CoV-2 cases before early December 2019.[6] However, available epidemiologic evidence strongly suggests that SARS-CoV-2 began infecting humans in Wuhan or the surrounding area between mid-October and early to mid-November 2019.[7]

While precedent of previous outbreaks of human infections from contact with animals favors the hypothesis that a natural zoonotic spillover is responsible for the origin of SARS-CoV-2, the emergence of SARS-CoV-2 that resulted in the COVID-19 pandemic was most likely the result of a research-related incident. This conclusion is not intended to be dispositive. The lack of transparency from government and public health officials in the PRC with respect to the origins of SARS-CoV-2 prevents reaching a more definitive conclusion. Should additional information be made publicly available, and subject to independent verification, it is possible that these conclusions would be subject to review and reconsideration.

Section I
Analysis of Natural Zoonotic Origins Hypothesis

Zoonotic spillovers, in which animal diseases cross the species barrier and infect humans, are a well-known, well-documented natural phenomena.[8] By some estimations, natural zoonotic spillovers are responsible for 60 to 75 percent of emerging diseases in humans.[9] Coronaviruses, to which SARS-CoV-2 belongs, are a large family of viruses that cause disease in a variety of domestic and farmed animals and have been responsible for previous outbreaks of new diseases in humans.[10] All coronaviruses known to infect humans are the result of natural zoonotic spillover from animals into humans.[11]

Two recent and prominent examples include Severe Acute Respiratory Syndrome ("SARS") and Middle East Respiratory Syndrome ("MERS"), both of which are caused by a coronavirus ("SARS-CoV" and "MERS-CoV" respectively) leading to severe respiratory disease in humans.[12] Moreover, recent infectious disease pandemics, with the exception of the 1977 Russian Flu pandemic, are believed to have natural zoonotic origins.[13]

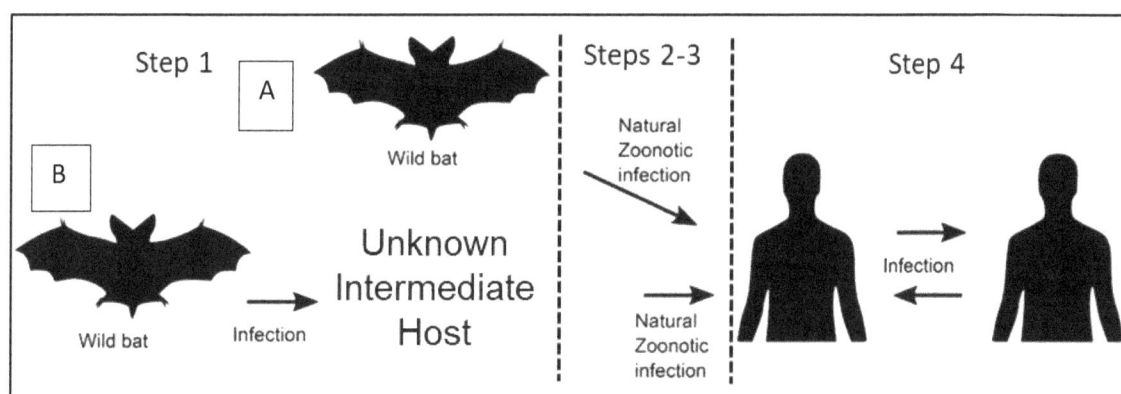

Figure 1: Example of a zoonotic spillover from bats. A: Direct spillover from bat to humans followed by human-to-human transmission. B: Spillover from a bat to an unknown intermediate host and then to humans, followed by human-to-human transmission.[14]

Natural zoonotic spillovers are a sequential process.[15] In this process, an animal virus must evolve in order to become a human-adapted virus. First, a virus infects animals. Second, those infected animals come into contact with humans (known as the human-animal interface). Third, the virus is able to infect humans. Fourth, the virus is able to adapt to efficiently transmit between humans.[16] Thus, a spillover event, in which disease is spread from animal to human, can result in one of two outcomes—either the pathogen, once transmitted from animals, is then transmitted from humans to humans, or the pathogen does not spread, resulting in a "dead-end" spillover. In many respects, once human-to-human transmission of SARS-CoV-2 was established, the onward human-to-human transmission of the virus would look similar regardless of whether it originated from a natural zoonotic spillover or a research-related incident.[17]

The natural zoonotic spillover hypothesis is a plausible explanation for how the COVID-19 pandemic started. There are a number of anomalies in the SARS-CoV-2 outbreak and the early COVID-19 pandemic compared to the emergence of past natural zoonotic spillovers, most notably the 2002-2004 SARS epidemic.

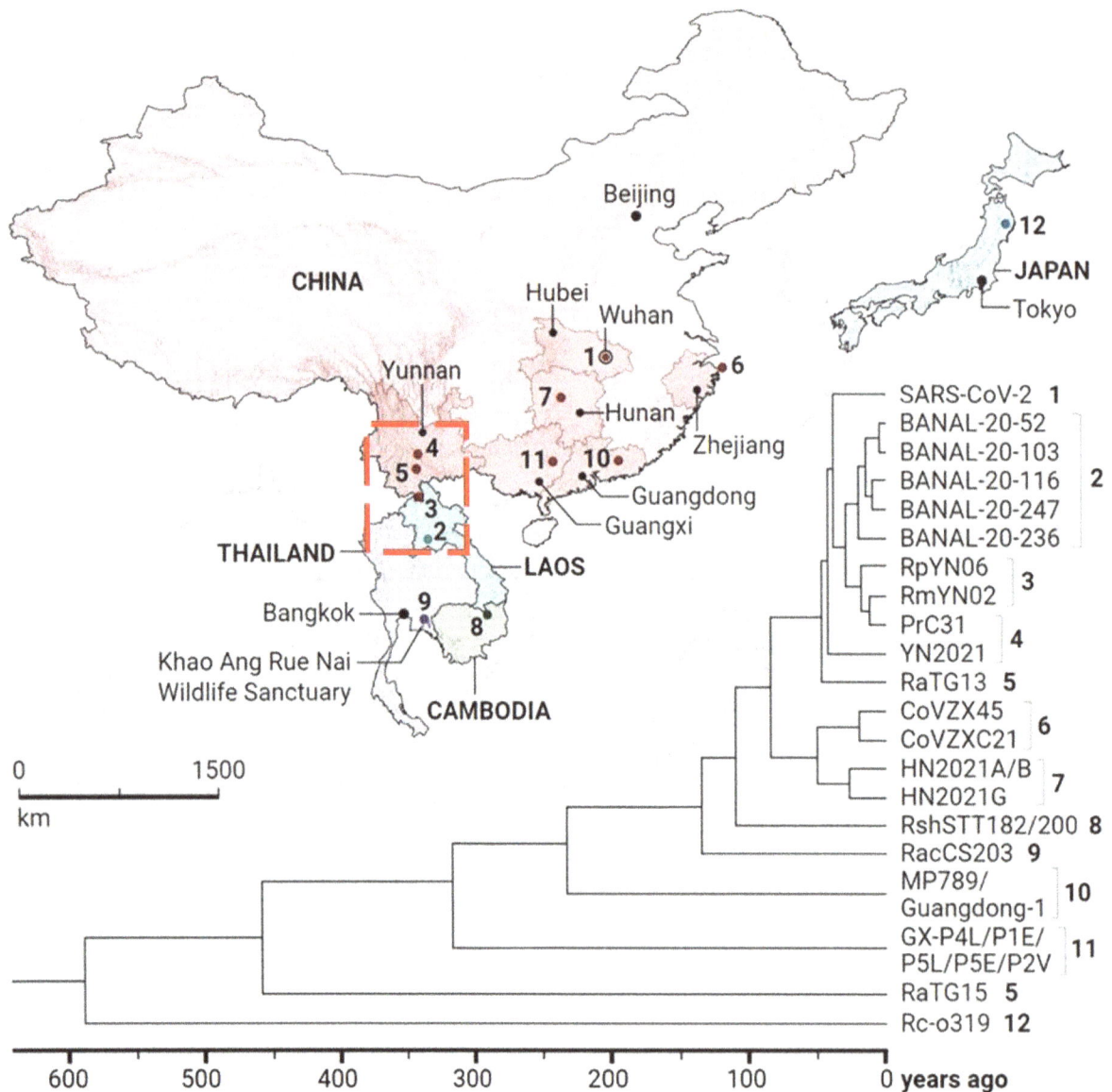

Figure 2: Map showing location of known SARS-related viruses most closely related to SARS-CoV-2 with five most closely related SARS-related coronaviruses to SARS-CoV-2 within the red box.[18]

Based on the precedent of past natural zoonotic spillovers, if SARS-CoV-2 is the result of a zoonotic spillover, it likely needed to circulate in an intermediate host to increase the virus' chances of being able to infect and replicate in humans.[19] Adaptation during circulation in an intermediate host is believed to have played a critical role in the emergence of SARS and MERS, as well as other bat viruses, such as hendra.[20] The identity of SARS-CoV-2's intermediate animal species remains unknown.[21] If such an intermediate animal species exists, where these intermediate species came into contact with and first infected humans is also unknown.[22] While it is likely that SARS-CoV-2 originated from a bat virus, most likely one found in horseshoe bats residing in Southern China or Southeast Asia, it remains unknown how SARS-CoV-2 traveled more than 1,000 miles from Southern China or Southeast Asia before emerging in Wuhan.[23] Almost three years after the COVID-19 pandemic began there is still no evidence of an animal

6

infected with SARS-CoV-2, or a closely related virus, before the first publicly reported human COVID-19 cases in Wuhan in December 2019.[24]

a. Epidemiology of SARS-CoV-2 Outbreak Differs from Previous Natural Zoonotic Spillovers

Most recent natural zoonotic spillovers of respiratory viruses with pandemic potential have left behind evidence of where and how they occurred.[25] Failed inter-species transmissions, or "dead-end" spillovers, typically leave behind serological evidence in the form of antibodies in humans and animals that were exposed and infected but did not effectively transmit the virus to others.[26] Failed transmissions also typically leave behind genetic evidence at the animal-human interface.[27]

Like interspecies transmission, human-to-human transmission also leave behind epidemiological evidence. The SARS epidemic saw at least five independent spillovers of the SARS virus into humans that then spread the virus to other humans, with other spillovers likely going unidentified and failing to cause sustained chains of transmission.[28] These spillovers occurred across multiple geographically distant live animal markets in Guangdong Province, China over a period of several months in 2002-2003.[29] Late-2003 to 2004 also saw isolated outbreaks of human SARS cases caused by additional independent spillovers of the virus.[30] Within six months of the start of the 2002-2004 SARS epidemic, intermediate host animal species candidates were identified, and numerous animals infected with SARS were found soon after the outbreak was identified.[31] In addition, early SARS virus samples retrieved from infected humans contained genetic mutations that reflected its period of circulation and adaptation in palm civets, the intermediate species.[32]

SARS-CoV-2's emergence also contrasts with outbreaks of human cases of Avian Influenza H7N9 in 2013. Like the 2002-2004 SARS outbreak, H7N9 started with multiple independent introductions of the virus into humans across multiple locations, even though the total number of human infections numbered less than 500.[33] Geographically disparate, independent spillovers imply that H7N9 Avian Influenza had circulated in bird populations for some time and across several provinces in China before the first known human infections. This is in contrast to the lack of geographically disparate cases of early COVID-19 cases in Hubei or China.[34]

The occurrence of natural zoonotic spillovers is also determined in part by probability. The frequency with which humans are exposed to an intermediate animal species infected with a zoonotic viral agent "is likely to be an important determinant in disease emergence." [35] This makes poorly regulated live animal markets in China and Southeast Asia effective conduits of zoonotic diseases.[36] The crowded conditions at these live animal markets mean that different members of multiple animal species that ordinarily would not come into contact are placed in close proximity to each other and large numbers of humans. These animals are often in poor health and shed viruses.[37]

7

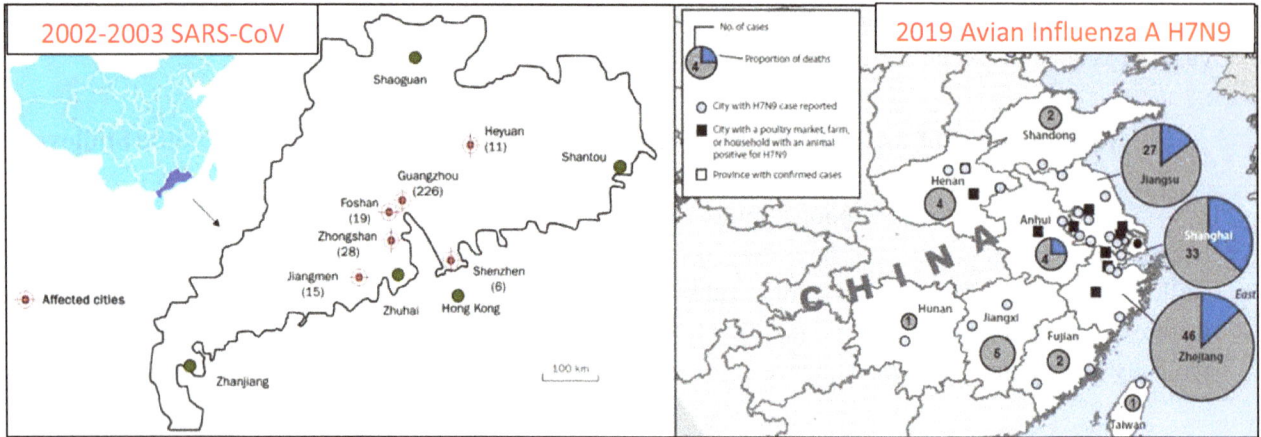

Figure 3: Comparison of Early Outbreaks of SARS-CoV and Avian Influenza H7N9.

Left: Map showing geographic distribution of SARS outbreak in Guangdong Province with dates of independent outbreaks of SARS from Nov. 2002 to Jan. 2003.[38]

Right: Map of confirmed human cases of avian influenza A (H7N9) from Feb. 19, 2013 to April 29, 2013.[39]

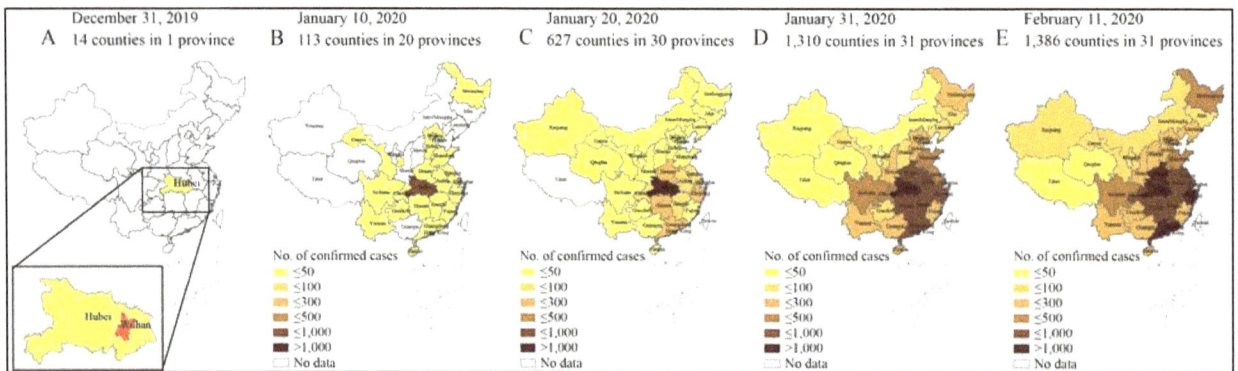

Figure 4: Map showing geo-temporal spread of COVID-19 in China from Dec. 31, 2019 to Feb. 11, 2020, starting only in Wuhan.[40]

A number of epidemiologists and virologists – and, at first, the Chinese government – have asserted that the COVID-19 pandemic originated from a natural zoonotic transmission occurring at the Huanan Seafood Market.[41] Government officials in China have subsequently also postulated the theory that SARS-CoV-2 arrived in China on the surface of imported frozen seafood or was brought into China by infected people or animals after being created by the U.S. military. Support for these alternative theories is limited to government-controlled publications in China and is not credible absent independent corroboration.[42]

Two key facts bolster the natural zoonotic origin argument. First, approximately 33 percent of the earliest known human COVID-19 cases (with symptom onset dates in mid- to late-December 2019) were associated with the Huanan Seafood Market in Wuhan.[43] Second, a number of animal species susceptible to SARS-CoV-2 were sold alive and in poor animal welfare conditions at the market.[44]

However, there is no published genetic evidence that SARS-CoV-2 was circulating in animals *prior* to the start of the COVID-19 pandemic.[45] Additionally, the genomes of early COVID-19 cases did not show genetic evidence, in the form of adaptive mutations that SARS-CoV-2 recently circulated in another animal species other than humans.[46] Moreover, the genetic similarity between the environmental samples and

8

human viral samples supports the likelihood that the virus found at the Huanan Seafood Market was shed by infected humans, rather than by infected animals.[47]

There also do not appear to have been subsequent spillovers of the virus that generated sustained transmission in humans, or any other independent spillovers of SARS-CoV-2, from the intermediate host animal(s) to humans since the pandemic started.[48] It is also noteworthy that the earliest variants of SARS-CoV-2 were well-adapted for human-to-human transmission.[49]

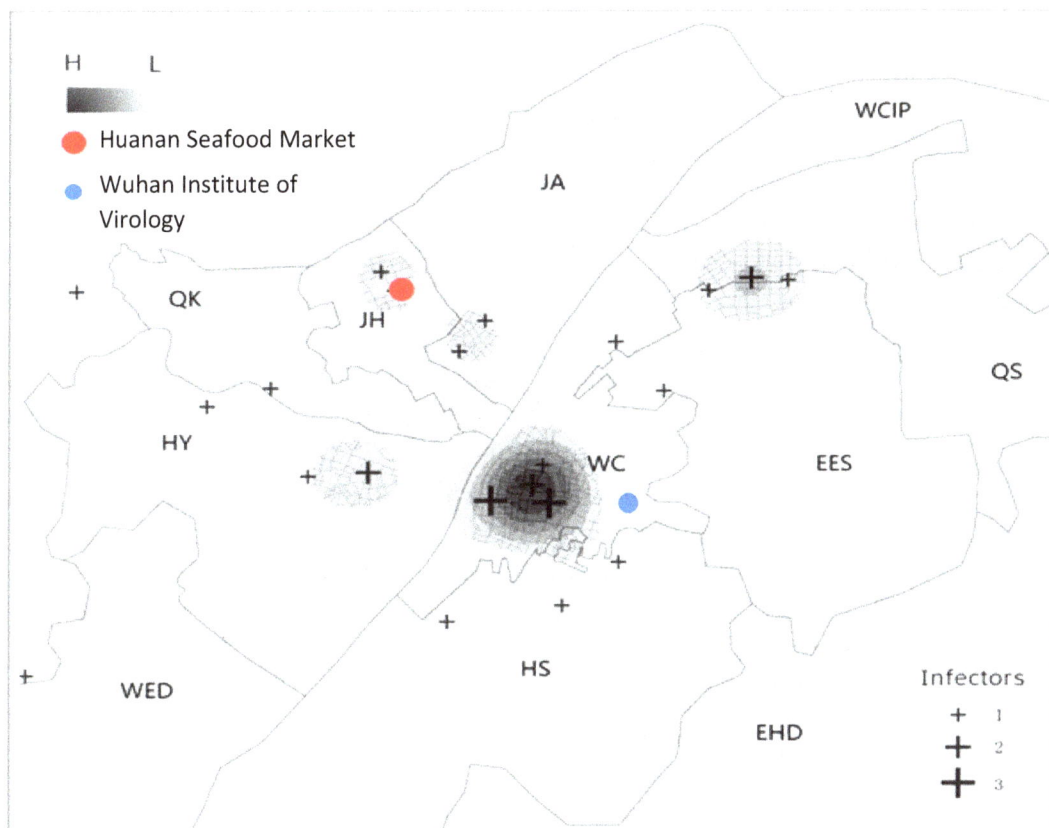

Figure 5: Spatial distribution of Weibo social media platform users who used COVID-19 assistance channel, a web application people searched when looking for flu-like symptoms, from Dec. 20, 2019 to Jan. 18, 2020, overlaid with location of Huanan Seafood Market and Wuhan Institute of Virology's campus in central Wuhan. (Adapted from: Peng, Z., Wang, R., Liu, L., & Wu, H. (2020). Exploring Urban Spatial Features of COVID-19 Transmission in Wuhan Baed on Social Media Data. ISPRS International Journal of Geo-Information, 9(6), 402. MDPI AG. Retrieved from http://dx.doi.org/10.3390/ijgi9060402).

These facts represent a significant break from the precedent of other zoonotic spillovers involving respiratory viruses, such as MERS and SARS. Relevant past zoonotic spillovers are those involving respiratory viruses that, like SARS-CoV-2, spread primarily through aerosols. Relatively recent spillovers involving live animal markets in urban areas are also relevant. Isolated spillovers of viruses in rural areas involving a small number of human infections have less precedential value, as do viruses that transmit primarily through close physical contact or are vector-borne. Accordingly, the SARS epidemic, the emergence of MERS, and several outbreaks of avian-influenzas have greater precedential value than viruses like monkeypox, Zika, human immunodeficiency virus (HIV), or Ebola, because the viruses and the circumstances of their emergence are more similar to that of SARS-CoV-2.

Early SARS-CoV-2 variants had little genetic diversity and were closely related to each other, differing by only two nucleotides out of approximately 29,900 nucleotides.[50] The fact that only two early variants of the virus have been identified indicates the virus had not been circulating widely or for a long period of time, and hence had little opportunity to mutate and cause new viral variants.[51] This also suggests that SARS-CoV-2 spilled over into humans only once or twice over an approximately two week period, and that these one to two spillovers resulted in sustained human-to-human transmission.[52] This successful spillover also only appears to have occurred in Wuhan or closely surrounding areas.[53]

Understanding the epidemiology of the outbreak is difficult, as the earliest known COVID-19 cases are unlikely to be the first humans actually infected with SARS-CoV-2.[54] The earliest identified COVID-19 cases, reported by PRC government officials, have a symptom onset date of December 8, 2019.[55] A majority of epidemiological modeling indicates that SARS-CoV-2 spilled over into humans between mid-October and early to mid-November 2019.[56] These early Wuhan cases seeded the virus in Wuhan as SARS-CoV-2 spread from person to person after the initial spillover event(s).[57]

The PRC has reported finding no retrospective evidence of COVID-19 cases in October or November 2019.[58] However, retrospective case searches by PRC public health authorities were limited to individuals requiring medical treatment.[59] As a result, the PRC's retrospective case search likely missed between 80 to 95 percent of all COVID-19 cases, which were asymptomatic or mildly symptomatic.[60] Undercounting of early COVID-19 cases is also likely due to China's restrictive case definitions which initially required not only severe COVID-19 symptoms, but a link to the Huanan Seafood Market.[61] It is estimated that during the period from mid-January to early March 2020, China's case definitions did not account for approximately 200,000 COVID-19 cases.[62]

b. Missing Evidence of a Natural Zoonotic Spillover

Environmental samples collected between January and March 2020 at the Huanan Seafood Market from countertops, fridges, gloves, and other surfaces tested positive for SARS-CoV-2.[63] According to presentations made to the World Health Organization (WHO) by PRC government officials and scientists in early 2020, none of the animals at the market when it was closed, in the market's supply chain, or in China's animal farming industry were infected with SARS-CoV-2.[64] That would be a significant variation from multiple precedents from previous natural zoonotic spillovers. For example, the discovery of infected palm civets during the SARS epidemic, and infected chickens and other farmed birds during multiple outbreaks of avian influenza, indicate a pattern where infected animals are expected to be present at the location of zoonotic spillovers and in the related supply chains.[65]

10

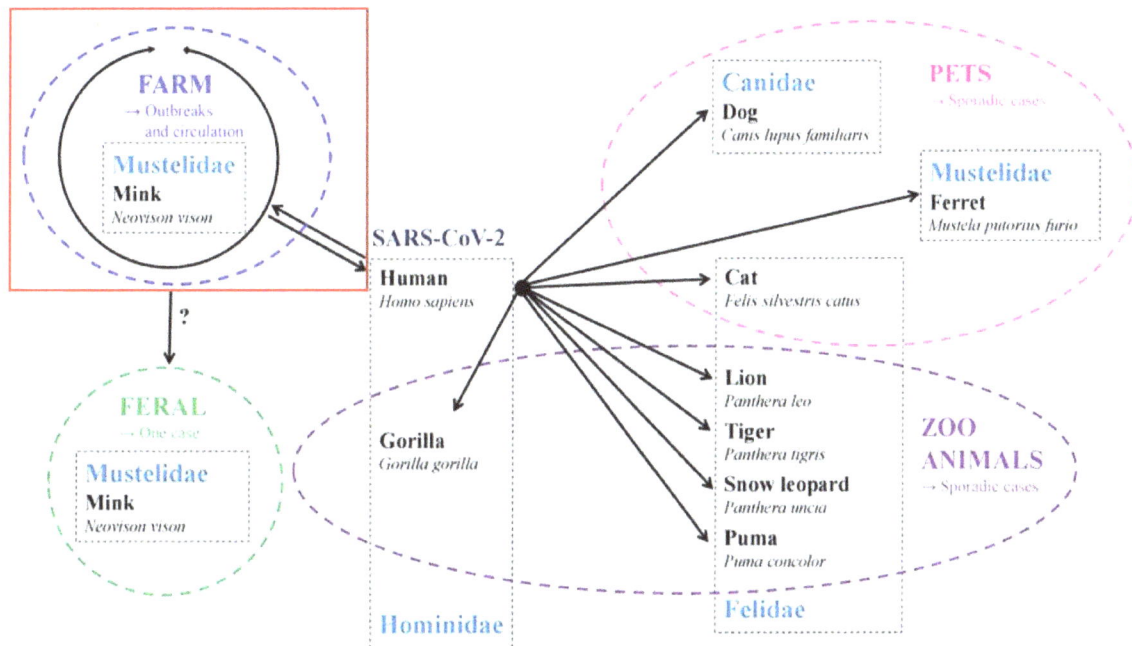

Figure 6: Animal species found to be naturally infected with SARS-CoV-2. Arrows show route of transmission. Red box highlights mink as the only animal known to have transmitted SARS-CoV-2 back to humans.[66]

Cases of human-to-animal transmission of SARS-CoV-2 have led to the identification of a number of mammal species susceptible to the virus that were sold at the Huanan Seafood Market, including mink, foxes, and raccoon dogs.[67] Of these, mink is the only industrially farmed animal identified to have transmitted SARS-CoV-2 from animals to humans with documented cases of farm workers being infected with mink-specific SARS-CoV-2 variants.[68,69]

China is the world's largest producer of farmed mink, raccoon dogs, and foxes.[70] Animal welfare conditions on these farms are poor and present an ideal environment for the spread and zoonotic spillover of SARS-CoV-2.[71] Scientists expect, because of SARS-CoV-2's ability to infect multiple species, that SARS-CoV-2 will likely become endemic in a number of wild animal populations, including mink, deer, and foxes.[72] However, PRC officials still have not reported a single SARS-CoV-2 infection in its farmed or wild mink, raccoon dog, or fox populations.[73] PRC officials and scientists have also reported to the WHO that they have not found a single instance of an animal infected with SARS-CoV-2 prior to the COVID-19 pandemic.[74]

c. **Problems with the Natural Zoonotic Hypothesis**

Based on precedent and genomics, the most likely scenario for a zoonotic origin of the COVID-19 pandemic is that SARS-CoV-2 crossed over the species barrier from an intermediate host to humans.[75] However, the available evidence is also consistent, perhaps more so, with a direct bat-to-human spillover. Both scenarios remain plausible and, in the absence of additional information, should be considered equally valid hypotheses.[76] However, nearly three years after the COVID-19 pandemic began, critical evidence that would prove that the emergence of SARS-CoV-2 and resulting COVID-19 pandemic was caused by a natural zoonotic spillover is missing.

As described in this report, the following facts and gaps in information are reasons why the natural zoonotic hypothesis is unlikely to explain the origins of SARS-CoV-2:

- The intermediate host species for SARS-CoV-2, if one exists, remains unidentified. By comparison, within six months of the first known human case of SARS, public health officials in China found SARS infections in palm civets and raccoon dogs in live animal markets in Guangdong Province.[77]

- Unlike SARS, the genomes of early COVID-19 cases from the first months of the pandemic do not show genetic evidence of SARS-CoV-2 having circulated in another animal species other than humans. None of the animals tested from the Huanan Seafood Market's supply chain, or in China's animal farming industry were infected with SARS-CoV-2, according to presentations by PRC officials to the WHO.[78]

- SARS-CoV-2's high binding affinity for human ACE2 receptors suggests that it is possible for it to directly infect humans without needing a period of adaptation in an intermediate host.[79] Direct spillover from a bat would explain the failure to find an intermediate host.[80] While direct bat-to-human spillover of coronaviruses has never been confirmed to cause a human outbreak, it is theoretically possible and there is circumstantial evidence suggesting it may occur under limited specific circumstances.[81]

- Based on the available evidence, Wuhan is the only location where SARS-CoV-2 spilled over into humans. [82] After the unidentified source transmitted SARS-CoV-2 to humans, it stopped transmitting SARS-CoV-2.[83] This is at odds with the precedent of the 2002-2004 SARS epidemic where infected palm civets continued to transmit the virus to humans and to raccoon dogs.[84] If the COVID-19 pandemic is the result of a zoonotic spillover from an intermediate host of SARS-CoV-2, the virus would be expected to continue to circulate in the infected intermediate host population, creating the potential for additional independent spillovers into humans and other animals.[85]

- The low genetic diversity of the earliest SARS-CoV-2 samples suggests that the COVID-19 pandemic is most likely the result of a single successful spillover of SARS-CoV-2.[86] Although the possibility of two spillover events cannot be ruled out, both the SARS epidemic and the 2013 avian influenza A (H7N9) outbreaks saw multiple independent spillovers of those viruses and exhibited much greater genetic diversity than early SARS-CoV-2 strains.

Based on this combination of factors, the available evidence appears to be inconsistent with both historic precedent and the scientific understanding of how natural zoonotic spillovers of respiratory viruses like SARS-CoV-2 occur. Ultimately, without increased transparency and publicly available and reproducible evidence that addresses these missing pieces of evidence, it is difficult to support the natural zoonotic origin hypothesis for the SARS-CoV-2 outbreak and COVID-19 pandemic.

Section II
Analysis of Research-Related Incident Hypothesis

Research-related incidents at labs in China, the United States, and elsewhere have happened and, in some instances, resulted in limited human-to-human transmission. For example, there have been at least six research related incidents involving the escape of SARS-CoV from high-containment laboratories in China (four), Taiwan (one), and Singapore (one).[87] The 1977 Influenza A (H1N1) pandemic is now widely accepted to have been the result of research-related incident, most likely a vaccine trial in the Soviet Union or China.[88] In June 2014, while investigating the unintentional exposure of one its researchers to potentially viable anthrax during an experiment in one of its biosafety level (BSL) 3 laboratories, the U.S. Centers for Disease Control and Prevention (CDC) discovered that a culture of non-pathogenic avian influenza was unintentionally cross-contaminated with the highly pathogenic H5N1 strain of influenza and shipped to a BSL3 U.S. Department of Agriculture laboratory.[89] There were no personnel exposures as a result of this event.

In short, human errors, mechanical failure, animal bites, animal escapes, inadequate training, insufficient funding, and pressure for results can lead to an escape of virulent pathogens, which could, in turn, infect animals and humans and lead to a release of a virus from a lab.

a. Coronavirus Research in Wuhan and at the Wuhan Institute of Virology

In the aftermath of the 2002-2004 SARS epidemic, Chinese authorities emphasized research on potential pandemic pathogens, including SARS-related coronaviruses, to develop vaccines and other medical countermeasures with the goal of attempting to predict and prevent the next coronavirus pandemic.[90] Wuhan is a global hub of coronavirus research. The Wuhan Institute of Virology is China's premier coronavirus research institute.[91] Although the WIV's coronavirus research is best documented because of its collaborations with western scientists, multiple institutions in Wuhan study coronaviruses including: Wuhan University, Huazhong Agricultural University, Hubei Centers for Disease Control and Prevention, Hubei Animal Centers for Disease Control and Prevention, Wuhan Centers for Disease Control and Prevention, and the Wuhan Institute of Biological Products, a vaccine manufacturing subsidiary of state-owned Sinopharm.[92] These institutes operate a number of biosafety level (BSL) 2, BSL3, and animal biosafety level (ABSL) 3 laboratories. Several of the BSL3 laboratories are relatively new, having been built only in the last five years. In all, laboratories are spread out across nine different campuses in Wuhan, with six hosting BSL3 or ABSL3 laboratories. The WIV is the only institute in Wuhan with a BSL4 laboratory.[93]

WIV researchers and their collaborators undertook large scale virus collection expeditions to Southern China and Southeast Asia, where bats naturally harbor SARS-related viruses, on an annual basis from 2004 onwards.[94] During these expeditions, scientists collected bat blood, saliva, and urine samples.[95] The WIV collected more than 15,000 bat-related samples around the time the pandemic began.[96] Of these, the WIV had identified more than 220 SARS-related coronaviruses, at least 100 of which have not been made public.[97]

13

Figure 7: Map of BSL2, 3, and 4 (including ABSL3) laboratories in Wuhan as of December 2019.[98]

WIV researchers actively sampled bats in Southern China and mainland Southeast Asia where the SARS-related coronaviruses most similar to SARS-CoV-2 have been collected and identified.[99] Viruses collected from these regions are 90.7 to 96.8 percent similar overall to SARS-CoV-2.[100] These include RaTG13, which was collected by WIV researchers in Yunnan Province.[101] RaTG13 is 96.3 percent genetically similar to SARS-CoV-2, and its existence was first made public only after the start of the COVID-19 pandemic, in February 2020.[102]

Presentations given by WIV researchers in 2018 show personnel on field expeditions wearing inadequate levels of personal protective equipment while handling bats.[103] Some personnel are photographed wearing "thin surgical masks and rubber gloves as they work [to collect bat samples], while others are unmasked with bare hands."[104] By contrast, a Wuhan Chinese Centers for Disease Control and Prevention (CCDC) scientist, who also regularly conducts bat sampling expeditions, said in a 2019 documentary that "[i]t is while discovering new viruses that we [researchers] are most at risk of infection."[105] The CCDC scientist further stated, "[i]f our skin is exposed, it can easily come in contact with bat excrement and contaminated matter, which means this is quite risky."[106]

Following field collection, samples were transported to Wuhan where they were screened for the presence of coronaviruses.[107] The WIV has two campuses, one in central Wuhan, Xiaohongshan, which

14

houses BSL2 and 3 laboratories, and a second, newer campus in Wuhan's southern suburbs, Zhengdian, which houses its BSL4 laboratory in addition to a BSL3 and multiple BSL2 laboratories. Researchers at the WIV then conducted experiments on newly isolated and sequenced coronaviruses.[108] Particular attention was given to SARS-related coronaviruses that have the ability to bind to human ACE2 receptors.[109] These viruses were considered by researchers at the WIV to be potential pandemic pathogens and pose a high-risk for spillover into humans.[110] Viruses were then sequenced and evaluated for their potential pandemic risk.[111]

The WIV conducted genetic recombination experiments as part of its coronavirus research in both BSL2 and BSL3 laboratories.[112] The WIV also conducted transgenic humanized mice experiments to assess the pandemic potential of SARS-related viruses.[113] They also tested the efficacy of vaccines in these mice and other animal species.[114] These animal experiments generate highly-infectious aerosols that are "ubiquitous... and are difficult to detect."[115] There were concerns about conducting this type of research in a BSL2 laboratory. As of May 2019, a Chinese CCDC biosafety expert expressed concern about China's lack of national BSL2 regulations, recommending that "[m]anipulation of highly pathogenic microorganisms should be performed in high level biosafety laboratories namely BSL3 or BSL4."[116]

This research process takes several years, leading to a multi-year gap between discovery of a virus and completing a paper ready for publication. For example, a virus genetically similar to SARS-CoV-2, the aforementioned RaTG13, was collected in 2013 and partially sequenced in 2016.[117] The remaining segments of RaTG13 were sequenced in 2018 and the sequence of the virus was finally made public in February 2020.[118] In another instance, one WIV graduate student took several years to publish data that resulted from field collection activities.[119]

b. **WIV Research on SARS-related Coronaviruses with Pandemic Potential**

By 2018, the WIV showed interest in finding SARS-related coronaviruses that used human ACE2 receptors to enter cells in order to determine whether SARS antibodies would effectively neutralize those viruses.[120] This research effort is described in a March 2018 grant proposal submitted to the Defense Advanced Research Projects Agency (DARPA) by a consortium of research entities, including the WIV, led by the U.S.-based non-governmental organization EcoHealth Alliance. The group proposed to collect and conduct genetic recombination experiments on SARS-related coronaviruses possessing specific traits making them "high-risk" for zoonotic spillover into animals and humans.[121]

Notably, the proposal describes the WIV's intent to search for SARS-related coronaviruses with potential to bind to human ACE2 receptors and that have naturally occurring furin cleavage sites in Yunnan Province, China.[122] According to the proposal, if WIV researchers were unable to find a SARS-related virus with these traits during sampling expeditions, they then proposed to manipulate the ACE2 receptors of SARS-related coronaviruses to increase binding affinity to human lung tissue and to insert furin cleavage sites at the same location where one appears in SARS-CoV-2.[123] This proposal was not ultimately funded by DARPA.

Furin cleavage sites are known to enhance virulence and increase viral replication in avian influenza and Ebola viruses. The grant proposal is in line with research trends in the field of virology in China. In 2015, researchers at Huazhong Agricultural University in Wuhan inserted an artificial furin cleavage site in Porcine Epidemic Diarrhea virus (an alpha coronavirus).[124] In 2019, researchers in China inserted a four amino acid furin cleavage site into Infectious Bronchitis coronavirus that affects poultry.[125] The WIV also received funding from PRC government agencies for research examining the spillover potential of SARS-related coronaviruses.[126]

In an interview with *Science*, Shi Zhengli, a senior scientist at the WIV and SARS-related coronavirus expert, disclosed that her team infected civets and mice that expressed human ACE2 receptors with chimeric SARS-related coronaviruses.[127] The results of these experiments indicated that SARS-related bat coronaviruses could infect and cause severe illness in humanized mice.[128] The WIV was later terminated as a sub-grantee by the National Institutes of Health (NIH) for failing to produce its laboratory notes and other records relating to these other experiments.[129]

c. WIV Biosafety and Biosecurity Patents and Procurements in 2019

Patents by WIV researchers published in 2018, 2019, and 2020, and procurements made by the WIV in 2019, indicate that the WIV struggled to maintain key biosafety capabilities at its high-containment BSL3 and BSL4 laboratories.[130] The following are examples of some of these patents and procurements:

- On April 24, 2019, WIV researchers submitted a patent for an auxiliary exhaust fan to maintain negative air pressure gradients in BSL3 and BSL4 high-containment laboratories.[131] This auxiliary fan was designed to prevent loss of negative pressure in the event of fan control failures, mechanical failures during fumigation, or human error.[132] These exhaust fans also addressed problems fumigating and disinfecting ventilation shafts and improving penetration of disinfectants into HEPA filters.[133]

- On August 14, 2019, the WIV issued a procurement notice for a project involving its environmental air disinfection system at the WIV's campus in central Wuhan.[134,135,136] The upgraded disinfection system used vaporized hydrogen peroxide to decontaminate laboratory surfaces.[137] A gaseous hydrogen peroxide disinfection system is an effective, less corrosive means to sterilize a laboratory than formaldehyde and other agents used by WIV researchers.[138]

- On September 16, 2019, the WIV issued a procurement notice seeking consultation for a "central air conditioning renovation project" at the new Zhengdian campus.[139] According to the U.S. CDC:

  > [HVAC] system design separates potentially contaminated laboratory air from areas outside the laboratory by maintaining the BSL-3/ABSL-3 areas at negative pressure to adjacent areas, by preventing re-circulation of laboratory exhaust air to other areas of the building, and by employing special engineering controls that prevent the occurrence of laboratory airflow reversals to outside the containment boundary.[140]

- On November 19, 2019, the WIV issued a sole source procurement request for an air incinerator at the original Xiaohongshan campus in central Wuhan.[141] The contract for the procurement was to be issued by December 5, 2019.[142] Air incinerators, though expensive to install and operate, were the mainstay of high-containment air sterilization prior to HEPA filtration. [143 , 144 , 145] The procurement stated that the incinerator was needed to sterilize exhaust gas from an autoclave, and that it would be added to the exhaust pipe after existing HEPA filters outside the autoclave to incinerate all the media discharged from within.[146]

- On December 11, 2019, WIV researchers filed a patent for a sensor to detect when biocontainment transfer cabinet HEPA filters had failed or were not operating correctly. [147] Experiments with infected animals often require moving the animals from a BSL3/BSL4 laboratory to a holding facility ABSL3/ABSL4 or transferring them from an animal holding room to a specific procedure room.[148] These animals create a variety of potentially hazardous infectious aerosols from urine, feces, fur, and by respiration.[149] The patent states, "when an accident occurs in the transportation process, an effective monitoring device is not available for judging whether the equipment is normal or not."[150]

- On November 13, 2020, WIV researchers filed a patent for a disinfectant formulation that improved upon one used in the Institute's high-containment laboratories. [151] The patented formulation "[r]educe[s] the corrosion effect to metal, especially stainless steel material."[152] As described in the patent, "[l]ong-term use [of the previous disinfectant] will lead to corrosion of metal components such as stainless steel, thereby reducing the protection of … facilities and equipment…shorten[s] its service life and cause economic losses, but also lead to the escape of highly pathogenic microorganisms into the external environment of the laboratory, resulting in loss of life and property and serious social problems."[153] The patent followed a March 2018 study that described WIV researchers using a disinfectant at a concentration more than three times higher than is recommended by the manufacturer.[154,155] The licensed U.S. manufacturer of the disinfectant states that "the higher … concentration, the more corrosive the solution will be."[156]

d. WIV Biosafety and Biosecurity Events in 2019

With the start of operations at the WIV's new BSL4 laboratory in late-2017 to 2018, government officials pressured WIV researchers to "leapfrog development" by conducting cutting-edge infectious disease research that contributed to China's national goals for biotechnology.[157] Throughout 2019, WIV experts published on challenging biosafety and biosecurity conditions faced by high-containment laboratories in China, including the WIV.

In May 2019, the Director of the WIV BSL4 laboratory warned that in high-containment laboratories in China:

> **Maintenance cost[s] [are] generally neglected; several high-level BSLs have insufficient operational funds for routine yet vital processes.** Due to the limited resources, some BSL-3 laboratories run on extremely minimal operational costs or in some cases none at all…

Currently, most laboratories lack specialized biosafety managers and engineers. In such facilities, some of the skilled staff is composed by part-time researchers. **This makes it difficult to identify and mitigate potential safety hazards in facility and equipment operation early enough**. Nonetheless, biosafety awareness, professional knowledge, and operational skill training still need to be improved among laboratory personnel. (emphasis added)[158]

In July 2019, China's National People's Congress drafted legislation, which later became law, to strengthen the management of laboratories involved in pathogen research and improve adherence to national standards and requirements for biosafety. It specifies that:

[L]ow-level pathogenic microorganism laboratories **shall not engage in pathogenic microorganism experiments that should be conducted in high-level pathogenic microorganism laboratories**…High-level pathogenic microorganism laboratories engaging in experimental activities of highly pathogenic or suspected highly pathogenic microorganisms shall be approved by the health or agriculture and rural authorities at or above the provincial level. For pathogenic microorganisms that have not been discovered or have been eliminated…relevant experimental activities shall not be carried out without approval. (emphasis added)[159]

Efforts by the WIV to improve biosafety were hampered by what officials called the "stranglehold problem," which meant a lack of access to advanced foreign biosafety technologies and materials.[160] Leadership at the WIV emphasized during a June 2019 meeting with WIV officials that addressing the "stranglehold problem" was critical to "pushing forward the construction and… development of science and technology for the nation."[161] The WIV's limited access to key foreign biosafety technologies forced the researchers to develop biosafety methods and construct equipment to remedy shortfalls.[162]

In July 2019, WIV leadership led a series of internal meetings on problems of operations in management at the WIV. The deputy director of the BSL4 laboratory issued a report on biocontainment equipment shortages and the impact of meeting the research goals of the government.[163] The report cited major problems that existed in the BSL4 laboratory including "hardware and technological aspects of the laboratory facilities" and "the management of biosafety."[164] The same report noted that the Director of the WIV urged the institute's senior personnel to "prioritize solving the urgent problems we are currently facing."[165]

On September 12, 2019 between the hours of 2:00 and 3:00 a.m. local time,[166] the WIV took down its online depository of data on viral sequences called the Wildlife-Borne Viral Pathogen Database.[167] The database was intermittently accessible from December 2019 to February 2020, before being permanently taken offline February 2020.[168] This database was previously accessible to the public, with the exception of a password protected section, which held unpublished sequence data accessible only to WIV personnel.[169] The WIV had a collection of more than 15,000 samples from bats, from which they had identified more than 1,400 bat viruses, including an estimated 100 unpublished sequences of SARS-related coronaviruses – the genre of coronaviruses to which SARS-CoV-2 belongs.[170] More than three years after it was first disabled, public access to the database has not been restored.[171]

On November 12, 2019, the WIV's BSL4 laboratory team issued a report on the achievements of the BSL4 laboratory since it began operations in 2018.[172] With respect to the "stranglehold problem", the report states that the WIV had overcome "the three no's" of "no equipment and technology standards, no design and construction teams, and no experience operating or maintaining" a high-containment laboratory.[173] The report continues to say that WIV personnel "brought into reality the 'three haves' of a complete system of standards, a superior team that operates and maintains [the lab], and valuable experience with construction." [174] This was achieved by "reinventing" imported equipment to make "the lab construction satisfy domestic and international standards" and making the French design of the BSL4 laboratory "conform to the requirements of Chinese construction."[175]

The report also described a high-pressure work environment. "In the laboratory, they often need to work for four consecutive hours, even extending to six hours," the report revealed. "During this time, they cannot eat, drink, or relieve themselves. This is an extreme test of a person's will and physical endurance. This not only demands that research personnel possess proficient operational skills, but they must also possess the ability to respond to various unexpected situations."[176]

The November 12, 2019 report suggested a biosafety problem had occurred at the WIV sometime before November 2019:

> Owing to [the fact] that the subject of research at the P4 lab is highly pathogenic microorganisms, inside the laboratory, once you have opened the stored test tubes, it is just as if having opened Pandora's Box. These viruses come without a shadow and leave without a trace. Although [we have] various preventive and protective measures, it is nevertheless necessary for lab personnel to operate very cautiously to avoid operational errors that give rise to dangers. **Every time this has happened, the members of the Zhengdian Lab [BSL4] Party Branch have always run to the frontline, and they have taken real action to mobilize and motivate other research personnel.** (emphasis added)[177]

On November 19, 2019, seven days after the BSL4 teams' report was issued, the WIV hosted a special training session run by a senior Chinese Academy of Sciences biosafety/biosecurity official who relayed "important oral and written instructions" from PRC leadership in Beijing to the WIV regarding the "complex and grave situation facing [bio]security work."[178] At the same training session, the Deputy Director of the Office of Safety and Security at the WIV "pointed to the severe consequences that could result from hidden safety dangers, and stressed that the rectification of hidden safety risks must be thorough, and management standards must be maintained."[179]

Section III
China's Early COVID-19 Vaccine Development Versus the U.S. Operation Warp Speed

Once the scale of the COVID-19 pandemic became clear, governments around the world scrambled to accelerate development of a vaccine to prevent death and severe disease from infection. In order to start vaccine development, researchers required the complete sequence of the target virus.[180] The full genetic sequence of SARS-CoV-2 was first posted to a global virus database on January 11, 2020 by a professor in China who acted in violation of PRC government restrictions on sharing information about SARS-CoV-2. As a consequence of his action, his laboratory was shut down for "rectification."[181]

After the SARS-CoV-2's sequence became available, vaccine developers inserted portions of the viral sequence into cells to produce the proteins that elicit an immune system response.[182] The cells that produce the proteins are called "constructs" and have to be created before vaccine development can begin.[183] After the construct is complete, the next developmental steps are preclinical animal toxicity, safety and efficacy studies, human clinical safety and efficacy trials, and commercial scale vaccine production.[184] Typically, these steps are done sequentially.[185]

During the COVID-19 pandemic, the urgent need for a vaccine resulted in these steps being done concurrently, which reduced the time spent on each step from years to a few months.[186] However, while pre-clinical studies and vaccine production can be done simultaneously, each step has its own timeline to completion that is difficult to compress. For example, animal studies are designed to last a specific length of time and cannot be curtailed without compromising the resulting data.[187] Similarly, the time it takes to grow the amount of vaccine needed for phase I trials is a limiting step, depending on the vaccine platform and scale of production.

a. U.S. Operation Warp Speed

The companies with candidate vaccines that would later be funded and supported by Operation Warp Speed in the United States all started vaccine development work on January 11, 2020 after the public release of the first SARS-CoV-2 sequence.[188] While mRNA vaccine candidates were able to design their vaccine construct in two days, because mRNA vaccines only need the coronavirus' genetic sequence to make a vaccine and no virus has to be cultivated in labs, traditional vaccine platforms take longer.[189]

The fastest of the Operation Warp Speed vaccine candidates to enter phase I human clinical trials among the non-mRNA vaccines was AstraZenca-Oxford's vaccine, ChAdOx1.[190] The AstraZeneca-Oxford team leveraged an existing vaccine construct and extensive experience with it to advance their candidate into phase I human clinical trials in an unprecedented 103 days.[191] Johnson & Johnson's vaccine candidate, Ad26, went from sequence to phase I clinical trials in 185 days.[192] As with AstraZeneca-Oxford, Johnson & Johnson was able to modify an existing construct it had developed for Ebola, as well as extensive institutional experience in vaccine development.[193] Both Ad26 and ChAdOx1 were adenovirus vaccines, in which a weakened version of the virus that cannot replicate is used to stimulate an immune reaction.[194]

Operation Warp Speed brought the first COVID-19 vaccines from sequence publication to regulatory approval in approximately eight months; "[o]ther medical miracles have been achieved, but few with the speed and success of developing the Covid-19 vaccines."[195] Operation Warp Speed accelerated development of COVID-19 vaccines by coordinating with the U.S. Food and Drug Administration (FDA) and Centers for Disease Control and Prevention, providing technical assistance, breaking through supply chain and manufacturing bottlenecks with the Defense Production Act, and de-risking vaccine development through guaranteed purchase agreements.[196] Vaccine developers ran clinical trials concurrently and on an accelerated timeline. The lessons learned from Operation Warp Speed have been widely shared, studied, and publicized, so it can serve as a model for how to quickly mobilize the government and private sector in response to an emergency.[197]

b. China's COVID-19 Vaccine Development Program

China also initiated a COVID-19 vaccine development with at least four research teams involved.[198] China did not initially have a mRNA vaccine candidate.[199] Two of these research teams were from the People's Liberation Army's Academy of Military Medical Sciences (AMMS), with the others from the Chinese Academy of Sciences (CAS) and the Chinese Centers for Disease Control and Prevention (CCDC).[200] The two AMMS teams reached notable early milestones in COVID-19 vaccine development. One AMMS team, led by Major General Chen Wei, using the same adenovirus vaccine platform as AstraZeneca-Oxford and Johnson & Johnson, went from sequence publication on January 11, 2020 to phase I human clinical trials on March 18, 2020, a span of only 67 days.[201]

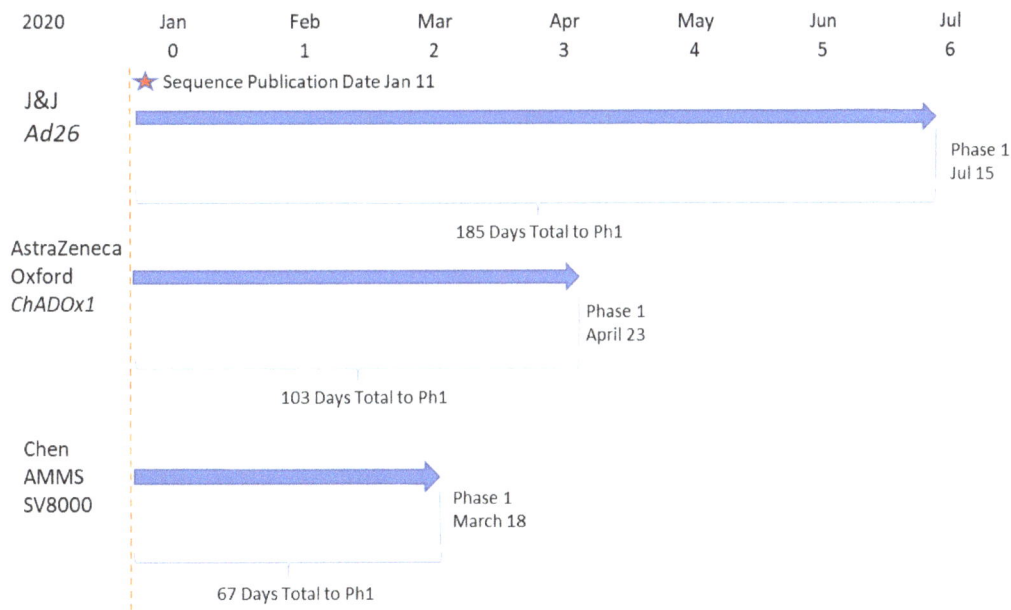

Figure 8: Comparison of Adenovirus Platform Timelines. Operation Warp Speed Vaccines: Johnson & Johnson's Ad26 and AstraZeneca-Oxford's ChADOx1 compared to Chen-AMMS's SV8000

The second AMMS team, led by Brigadier General Yusen Zhou, was the first to patent a COVID-19 vaccine on February 24, 2020.[202] The Zhou AMMS team's patent included data from a mouse experiment showing that the vaccine construct neutralized SARS-CoV-2 infections.[203] Other researchers in China working with the same vaccine platform took between three to four months to develop their candidate

21

vaccine.[204] The Zhou AMMS COVID-19 vaccine candidate does not appear to have advanced into phase I human clinical trials.[205] The Chen AMMS COVID-19 vaccine is commercially produced by CanSino.[206]

Given Operation Warp Speed's success, it is unusual that the two AMMS COVID-19 vaccine development teams were able to reach early milestones in vaccine development even more quickly. The Chen AMMS team beat AstraZeneca-Oxford to phase I clinical trials by 38 days. The Zhou AMMS team built and validated the effectiveness of its COVID-19 candidate vaccine 44 days after the sequence of SARS-CoV-2 was released. The extremely accelerated vaccines development timelines achieved by the AMMS teams pose the following two outstanding questions:

- What additional steps, processes, or novel techniques did AMMS researchers take that advanced the development of their vaccine faster than the Operation Warp Speed timeline?

- If no additional steps were taken to speed up the development timeline, when did researchers in China have access to the genomic sequence? Was it before January 11, 2020? If so, how far in advance of January 11, 2020?

Section IV

Basis for Assessment that Research-Related Incident is More Likely Origin of SARS-CoV-2

Nearly three years after the COVID-19 pandemic began, substantial evidence demonstrating that the COVID-19 pandemic was the result of a research-related incident has emerged. A research-related incident is consistent with the early epidemiology showing rapid spread of the virus in Wuhan, with the earliest calls for assistance being located in the near the WIV's original campus in central Wuhan.[207] It also explains the low genetic diversity of the earliest known SARS-CoV-2 human infections in Wuhan, because the likely index case, would be an infected researcher, is the likely primary source of the virus in Wuhan.[208] A research-related incident also explains the failure to find an intermediate host as well as the failure to find any animal infections pre-dating human COVID-19 cases.[209]

Although the WIV's coronavirus research is best documented because of its collaborations with western scientists, multiple institutions in Wuhan study coronaviruses including: Wuhan University, Huazhong Agricultural University, Hubei Centers for Disease Control and Prevention, Hubei Animal Centers for Disease Control and Prevention, Wuhan Centers for Disease Control and Prevention, and the Wuhan Institute of Biological Products, a vaccine manufacturing subsidiary of state-owned Sinopharm.

a. Coronavirus Research at the Wuhan Institute of Virology

The WIV is an epicenter of advanced coronavirus research that was designed to predict and prevent future pandemics by collecting, characterizing, and experimenting on "high-risk" coronavirus with the potential to spill over into humans:

- In the aftermath of the 2002-2004 SARS epidemic, WIV researchers undertook annual virus collection expeditions to Southern China and Southeast Asia, where bats naturally harbor SARS-related viruses, from 2004 onward.[210]

- WIV researchers actively sampled bats in Southern China and Southeast Asia where the SARS-related coronaviruses most similar to SARS-CoV-2 have been collected and identified.[211]

- The WIV had collected more than 15,000 samples from bats, from which they had identified more than 1,400 bat viruses, including an estimated 100 unpublished sequences of SARS-related coronaviruses – the genre of coronaviruses to which SARS-CoV-2 belongs.[212] The database containing the sequences of viruses collected by the WIV, including unpublished SARS-related coronaviruses, was taken offline starting in September 2019.

- Following field collection, samples were transported to Wuhan, where they were screened for the presence of coronaviruses.[213] WIV researchers performed animal and human cell-related research using recombinant genetic techniques with the express goal of discovering human adapted SARS-like chimeric viruses. The WIV conducted these experiments in BSL2 and BSL3 laboratories.

- Senior coronavirus researcher Shi Zhengli disclosed that in 2018-2020, her team infected civets and humanized mice that expressed human ACE2 receptors with chimeric SARS-related coronaviruses.[214] The results of these experiments have never been published.

- The EcoHealth Alliance NIH grants and DARPA grant proposals, in partnership with the WIV, sought to collect and conduct genetic recombination experiments on SARS-related coronaviruses with specific traits that made those viruses a "high-risk" for zoonotic spillover into animals and humans.[215] SARS-CoV-2 shares many of the traits these researchers were interested in finding in SARS-related coronaviruses or interested in engineering such traits if they were not found naturally.

b. Evidence of Biosafety Failures at the WIV

WIV patents and procurements suggest that the WIV experienced persistent biosafety problems relevant to the containment of an aerosolized respiratory virus like SARS-CoV-2.

- April 24, 2019:   Auxiliary exhaust patent
- August 14, 2019:   Environmental air disinfection system procurement
- September 16, 2019:   Central air conditioning
- November 19, 2019:   Sole source procurement for air incinerator
- December 11, 2019:   Biocontainment transfer cabinet HEPA filter failure patent
- November 13, 2020:   Disinfectant formulation patent

c. Management and training concerns at the WIV

Academic articles, reports, and meetings from the WIV also suggest that the WIV experienced persistent biosafety problems relevant to the containment of an aerosolized respiratory virus like SARS-CoV-2:

- In May 2019, the Director of the WIV BSL4 laboratory warned that in high-containment laboratories in China maintenance costs were neglected and part-time researchers made it **"difficult to identify and mitigate potential safety hazards in facility and equipment operation early enough."** (emphasis added) [216]

- Leadership at the WIV emphasized during a June 2019 meeting with WIV officials that addressing the "stranglehold problem" was critical to "pushing forward the construction and... development of science and technology for the nation."[217]

- In July 2019, the deputy director of the BSL4 laboratory issued a report on shortages of biosafety equipment and its impact on meeting the research expectations of the government.[218]

- In July 2019, China's National People's Congress began the process of drafting the law to strengthen the management of laboratories involved in pathogen research and improve adherence to national standards and requirements for biosafety.[219]

- A November 12, 2019 report suggested a biosafety problem had occurred at the WIV sometime before November 2019.[220]

- On November 19, 2019, the WIV hosted a special training session by the senior Chinese Academy of Sciences biosafety/biosecurity official who relayed "important oral and written instructions" from PRC leadership to the WIV regarding the "complex and grave situation facing [bio]security work."[221] This one-day training session for senior leadership was followed on November 20-21, 2019 with two days of safety training for personnel from the WIV and other Wuhan area high-containment laboratories.

d. Anomalies in Epidemiology of SARS-COV-2 Outbreak

- SARS-CoV-2 spilled over into humans only in Wuhan.[222] This is a break with the precedent of SARS, MERS, and multiple outbreaks of avian influenza, all of which were much less transmissible than SARS-CoV-2 and infected fewer animals.

- The low genetic diversity of the earliest SARS-CoV-2 samples, coupled with one of the two early lineages being more closely related to bat coronaviruses, suggests that COVID-19 pandemic is most likely the result of one, or at most two, spillovers of SARS-CoV-2.[223] SARS-CoV-2's low initial genetic diversity is also a break with the precedent of recent zoonotic spillovers of respiratory viruses.

- Critical corroborating evidence of a natural zoonotic spillover is missing. While the absence of evidence is not itself evidence, the lack of corroborating evidence of a zoonotic spillover or spillovers, three years into the pandemic, is highly problematic. If the COVID-19 pandemic is the result of the zoonotic spillover of SARS-CoV-2 in Wuhan from an intermediate host species, there should be evidence of SARS-CoV-2 circulating in animals before it spilled over into humans. Instead, there is no evidence that any animal was infected with SARS-CoV-2 prior to the first human cases.[224]

Conclusion

As noted by the WHO Scientific Advisory Group for the Origins of Novel Pathogens, the COVID-19 Lancet Commission, and the U.S. Office of the Director of National Intelligence 90-Day Assessment on the COVID-19 Origins, more information is needed to arrive at a more precise, if not a definitive, understanding of the origins of SARS-CoV-2 and how the COVID-19 pandemic began.[225] Governments, leaders, public health officials, and scientists involved in addressing the COVID-19 pandemic and working to prevent future pandemics, must commit to greater transparency, engagement, and responsibility in their efforts.

Based on the analysis of the publicly available information, it appears reasonable to conclude that the COVID-19 pandemic was, more likely than not, the result of a research-related incident. New information, made publicly available and independently verifiable, could change this assessment. However, the hypothesis of a natural zoonotic origin no longer deserves the benefit of the doubt, or the presumption of accuracy. The following are critical outstanding questions that would need to be addressed to be able to more definitively conclude the origins of SARS-CoV-2:

- What is the intermediate host species for SARS-CoV-2? Where did it first infect humans?
- Where is SARS-CoV-2's viral reservoir?
- How did SARS-CoV-2 acquire its unique genetic features, such as its furin cleavage site?

Advocates of a zoonotic origin theory must provide clear and convincing evidence that a natural zoonotic spillover is the source of the pandemic, as was demonstrated for the 2002-2004 SARS outbreak. In other words, there needs to be verifiable evidence that a natural zoonotic spillover actually occurred, not simply that such a spillover could have occurred.

[1] Tan, C. C. S., Lam, S. D., Richard, D., Owen, C. J., Berchtold, D., Orengo, C., Nair, M. S., Kuchipudi, S. V., Kapur, V., van Dorp, L., & Balloux, F. (2022). Transmission of SARS-CoV-2 from humans to animals and potential host adaptation. Nature Communications, 13(1). https://doi.org/10.1038/s41467-022-30698-6.

[2] Scientific Advisory Group for the Origins of Novel Pathogens (SAGO). (June 9, 202). Preliminary Report. World Health Organization. https://cdn.who.int/media/docs/default-source/scientific-advisory-group-on-the-origins-of-novel-pathogens/sago-report-09062022.pdf.

[3] *Id.*

[4] *Id.*

[5] China delayed releasing coronavirus info, frustrating WHO. (n.d.). AP NEWS. https://apnews.com/article/united-nations-health-ap-top-news-virus-outbreak-public-health-3c061794970661042b18d5aeaaed9fae.

[6] Cohen, Jon. (Aug. 18, 2022). Where did the pandemic start? Anywhere but here, argue papers by Chinese scientists echoing party line. Science. 2022: 377 (6608). https://www.science.org/content/article/pandemic-start-anywhere-but-here-argue-papers-chinese-scientists-echoing-party-line.

[7] Pekar J, Worobey M, Moshiri N, Scheffler K, Wertheim JO., et, al. (Mar. 18, 2021). Timing the SARS-CoV-2 index case in Hubei province. Science. 2021;372(6540):412-417. doi:10.1126/science.abf8003.

[8] Ellwanger JH, Chies JAB. (June 4, 2021). Zoonotic spillover: Understanding basic aspects for better prevention. Genet Mol Biol. 2021:44(1 Suppl 1). doi:10.1590/1678-4685-GMB-2020-0355.

[9] *Id.*

[10] *Id.*

[11] Ye ZW, Yuan S, Yuen KS, Fung SY, Chan CP, Jin DY. (Mar. 15, 2020). Zoonotic origins of human coronaviruses. Int J Biol Sci. 2020:16(10):1686-1697. doi:10.7150/ijbs.45472.

[12]*Id.*

[13] Rozo M, Gronvall GK. (Aug. 18, 2015).The Reemergent 1977 H1N1 Strain and the Gain-of-Function Debate. mBio. 2015;6(4):e01013-15.. doi:10.1128/mBio.01013-15; Pike BL, Saylors KE, Fair JN, et al. (June 2010). The Origin and Prevention of Pandemics. Clin Infect Dis. 2010;50(12):1636-1640. doi:10.1086/652860.

[14] Adapted from Segreto R, Deigin Y, McCairn K, Sousa A, Sirotkin D, Sirotkin K, Couey JJ, Jones A, Zhang D. (Mar. 25, 2021). Should we discount the laboratory origin of COVID-19? Environ Chem Lett. 2021;19(4):2743-2757. doi: 10.1007/s10311-021-01211-0.

[15] Plowright RK, Parrish CR, McCallum H, et al. (May 2017). Pathways to Zoonotic Spillover. Nature Reviews Microbiology. 2017;15(8):502-510. doi:10.1038/nrmicro.2017.45/

[16] *Id.*

[17] Worobey M, Levy JI, Malpica Serrano L, et al. (July 26, 2022).The Huanan Seafood Wholesale Market in Wuhan was the early epicenter of the COVID-19 pandemic. Science. 2022;377(6609):951-959. doi:10.1126/science.abp8715.

[18] Cohen, Jon. (April 22, 2022). Looking for Trouble. Science. 2022: 376 (6590). 10.1126/science.abq2305.

[19] *Supra*, note 17.

[20] *Supra*, note 8.

[21] Farag EA, Islam MM, Enan K, El-Hussein AM, Bansal D, Haroun M. SARS-CoV-2 at the human-animal interphase: A review. Heliyon. 2021;7(12):e08496. doi:10.1016/j.heliyon.2021.e08496

[22] Wang Q, Chen H, Shi Y, et al. (Sept. 29, 2021). Tracing the origins of SARS-CoV-2: lessons learned from the past. Cell Research. 31, 1139–1141. https://doi.org/10.1038/s41422-021-00575

[23] Lytras S, Xia W, Hughes J, Jiang X, Robertson DL. (Aug. 17. 2021). The animal origin of SARS-CoV-2. Science. 373(6558):968-970. doi:10.1126/science.abh0117.

[24] *Supra*, note 2.

[25] Ye ZW, Yuan S, Yuen KS, Fung SY, Chan CP, Jin DY. (Mar. 15, 2021). Zoonotic origins of human coronaviruses. Int J Biol Sci. 16(10):1686-1697. doi:10.7150/ijbs.45472.

[26] James M. Hughes, Mary E. Wilson, Brian L. Pike, Karen E. Saylors, Joseph N. Fair, Matthew LeBreton, Ubald Tamoufe, Cyrille F. Djoko, Anne W. Rimoin, Nathan D. Wolfe. (June 15, 2010). The Origin and Prevention of Pandemics. Clin Infect Dis. 2010. 50(12):1636-1640. doi:10.1086/652860/.

[27] *Id.*

[28] Wang, L.F., Eaton, B.T. (2007). Bats, Civets and the Emergence of SARS. In: Childs, J.E., Mackenzie, J.S., Richt, J.A. (eds) Wildlife and Emerging Zoonotic Diseases: The Biology, Circumstances and Consequences of Cross-Species Transmission. Current Topics in Microbiology and Immunology, vol 315. Springer, Berlin, Heidelberg. https://doi.org/10.1007/978-3-540-70962-6_13.

[29] *Id.*

[30] Liang G, Chen Q, Xu J, et al. (Oct. 10, 2004). Laboratory Diagnosis of Four Recent Sporadic Cases of Community-Acquired SARS, Guangdong Province, China. Emerg Infect Dis. https://doi.org/10.3201%2Feid1010.040445.

[31] *Supra*, note 28.

[32] *Id.*

[33] Butler D. (April 24, 2013). Mapping the H7N9 Avian Flu Outbreaks. Nature. https://doi.org/10.1038/nature.2013.12863.

[34] Jernigan, Daniel, et. al. (May 10, 2013). Emergence of Avian Influenza A(H7N9) Virus Causing Severe Human Illness — China, February–April 2013. Morbidity and Mortality Weekly Report (MMWR). Retrieved October 26, 2022, from https://www.cdc.gov/mmwr/preview/mmwrhtml/mm6218a6.htm#:~:text=April%2029%2C%202013 .

[35] *Supra*, note 26.

[36] *Supra*, note 16.

[37] *Supra*, note 8.

[38] Zhong N, Zheng B, Li Y, et al. (Oct. 25, 2003). Epidemiology and cause of severe acute respiratory syndrome (SARS) in Guangdong, People's Republic of China, in February, 2003. The Lancet. 362(9393):1353-1358. doi:10.1016/s0140-6736(03)14630-2).

[39] *Supra*, note 34.

[40] CDC Weekly, C. (2020). The Epidemiological Characteristics of an Outbreak of 2019 Novel Coronavirus Diseases (COVID-19) — China, 2020. China CDC Weekly. 2(8), 113–122. https://doi.org/10.46234/ccdcw2020.032.

[41] *Supra*, note 6.

[42] *Supra*, note 3.

[43] World Health Organization. (2021) "WHO-convened global study of origins of SARS-CoV-2: China Part"; https://www.who.int/publications/i/item/who-convened-global-study-of-origins-of-sars-cov-2-china-part.

[44] *Supra*, note 17.

[45] *Supra*, note 43.

[46] Zhan, Shing Hei & Deverman, Benjamin E. & Chan, Alina Yujia. (May 2, 2020). SARS-CoV-2 is well adapted for humans. What does this mean for re-emergence?. bioRxiv; doi: https://doi.org/10.1101/2020.05.01.073262.

[47] Gao, George & Liu, William & Liu, Peipei & Lei, Wenwen & Jia, Zhiyuan & He, Xiaozhou & Liu, Lin-Lin & Shi, Weifeng & Tan, Yun & Zou, Shumei & Zhao, Xiang & Wong, Gary & Wang, Ji & Wang, Feng & Wang, Gang & Qin, Kun & Gao, Rong-bao & Zhang, Jie & Li, Min & Wu, Guizhen. (Feb. 25, 2022). Surveillance of SARS-CoV-2 in the environment and animal samples of the Huanan Seafood Market. Research Square. https://doi.org/10.21203/rs.3.rs-1370392/v1.

[48] Pekar JE, Magee A, Parker E, et al. (Jul. 26, 2022). The Molecular Epidemiology of Multiple Zoonotic Origins of SARS-CoV-2. Science. 377(6609):960-966. https://doi:10.1126/science.abp8337.

[49] *Id.*

[50] *Supra*, note 43.

[51] *Supra*, note 48.

[52] *Id.*

[53] *Supra*, note 43.

[54] *Id.*

[55] *Id.*

[56] *Id.*

[57] Epidemiology Team. (Feb. 17, 2020). The epidemiological characteristics of an outbreak of 2019 novel coronavirus diseases (COVID-19)—China, 2020. China CDC weekly. 2(8); (2020): 113-122. doi: 10.46234/ccdcw2020.032

[58] *Supra*, note 43.

[59] *Id.*

[60] Menachemi N, Dixon BE, Wools-Kaloustian KK, Yiannoutsos CT, Halverson PK. (May-Jun. 2021). How Many SARS-CoV-2-Infected People Require Hospitalization? Using Random Sample Testing to Better Inform Preparedness Efforts. J Public Health Manag Pract. 01;27(3):246-250. doi: 10.1097/PHH.0000000000001331. PMID: 33729203; see also Wang, Vivian (Feb. 27, 2020). Most Coronavirus Cases Are Mild. That's Good and Bad News. The New York Times. https://www.nytimes.com/2020/02/27/world/asia/coronavirus-treament-recovery.html.

[61] *Id.*

[62] Tsang, Tim K., Peng Wu, Yun Lin, Eric HY Lau, Gabriel M. Leung, and Benjamin J. Cowling. (May 1, 2020). Effect of Changing Case Definitions for COVID-19 on the Epidemic Curve and Transmission Parameters in Mainland China: a Modelling Study. The Lancet Public Health. Vol.5, no. 5. https://doi.org/10.1016/S2468-2667(20)30089-X.

[63] *Supra*, note 43.

[64] *Supra,* note 2.

[65] Li, K., Guan, Y., Wang, J. et al. (Jul. 8, 2004). Genesis of a Highly Pathogenic and Potentially Pandemic H5N1 Influenza Virus in Eastern Asia. Nature. 430: 209–213. https://doi.org/10.1038/nature02746.

[66] Figure adapted from Fenollar F, Mediannikov O, Maurin M, Devaux C, Colson P, Levasseur A, Fournier P-E and Raoult D (April 1, 2021) Mink, SARS-CoV-2, and the Human Animal Interface. Front. Microbiol. 12:663815 https://doi:10.3389/fmicb.2021.663815.

[67] *Supra*, note 43.

[68] *Supra*, note 66.

[69] Pomorska-Mól M, Włodarek J, Gogulski M, Rybska M. (Jul. 15, 2021). Review: SARS-CoV-2 infection in farmed minks - an overview of current knowledge on occurrence, disease and epidemiology. Animal. 15(7):100272. https://doi.org/10.1016/j.animal.2021.100272

[70] Lung, Yuan-Chin & Lin Sophie. (July 2019). China's Fur Trade and Its Position in the Global Fur Industry. Act Asia. https://www.actasia.org/wp-content/uploads/2019/10/China-Fur-Report-7.5.pdf.

[71] Shah, S., & Comrie, T. (Jan. 19, 2022). Animals That Infect Humans Are Scary. It's Worse When We Infect Them Back. The New York Times. https://www.nytimes.com/2022/01/19/magazine/spillback-animal-disease.html

[72] Phillips, N. (Feb. 16, 2021). The Coronavirus is Here to Stay — Here's What That Means. Nature, 590(7846), 382–384. https://doi.org/10.1038/d41586-021-00396-2.

[73] *Supra*, note 43.

[74] *Id.*

[75] *Supra*, note 17.

[76] *Supra*, note 43.

[77] Guan, Y., Zheng, B. J., He, Y. Q., Liu, X. L., Zhuang, Z. X., Cheung, C. L., Luo, S. W., Li, P. H., Zhang, L. J., Guan, Y. J., Butt, K. M., Wong, K. L., Chan, K. W., Lim, W., Shortridge, K. F., Yuen, K. Y., Peiris, J. S., & Poon, L. L. (2003). Isolation and Characterization of Viruses Related to the SARS Coronavirus from Animals in Southern China. Science. 302(5643), 276–278. https://doi.org/10.1126/science.1087139.

[78] *Supra*, note 43.

[79] *Id.*

[80] *Id.*

[81] Wang, N., Li, S. Y., Yang, X. L., Huang, H. M., Zhang, Y. J., Guo, H., Luo, C. M., Miller, M., Zhu, G., Chmura, A. A., Hagan, E., Zhou, J. H., Zhang, Y. Z., Wang, L. F., Daszak, P., & Shi, Z. L. (2018). Serological Evidence of Bat SARS-Related Coronavirus Infection in Humans, China. Virologica Sinica, 33(1), 104–107. https://doi.org/10.1007/s12250-018-0012-7.

[82] *Supra*, note 17.

[83] The Novel Coronavirus Pneumonia Emergency Response Epidemiology Team. The Epidemiological Characteristics of an Outbreak of 2019 Novel Coronavirus Diseases (COVID-19) — China, 2020[J]. China CDC Weekly, 2020, 2(8): 113-122. doi: 10.46234/ccdcw2020.032.

[84] *Supra*, note 28.

[85] *Id.*

[86] Rambaut, A., Holmes, E. C., O'Toole, Á., Hill, V., McCrone, J. T., Ruis, C., du Plessis, L., & Pybus, O. G. (Mar. 26, 2020). Origins of SARS-CoV-2. World Health Organization. https://apps.who.int/iris/bitstream/handle/10665/332197/WHO-2019-nCoV-FAQ-Virus_origin-2020.1-eng.pdf

[87] Senior K. (Nov. 3, 2003). Recent Singapore SARS case a Laboratory Accident. The Lancet. Infectious Diseases. 3(11), 679. https://doi.org/10.1016/S1473-3099(03)00815-6; *see also:* Walgate R. (Apr. 27, 2004). SARS Escaped Beijing lab Twice. Genome Biology. 4: spotlight-20040427-03. https://doi.org/10.1186/gb-spotlight-20040427-0; Taiwan: CIDRAP (December 17, 2003), Taiwanese SARS researcher infected. University of Minnesota. (Dec. 17, 2033). Taiwanese SARS Researcher Infected. CIDRAP. https://www.cidrap.umn.edu/news-perspective/2003/12/taiwanese-sars-researcher-infected.

[88] *Supra,* note 13.

[89] CDC Press Release. (January 1, 2016). U.S. Centers for Disease Control and Prevention. https://www.cdc.gov/media/releases/2014/p0711-lab-safety.html.

[90] Wuhan Institute of Virology. (n.d.). History-Wuhan Institute of Virology. institute.wuhanvirology.org. Accessed October 10, 2022. http://institute.wuhanvirology.org/About_Us2016/History2016/index.htm.

[91] BurNIH-00000483-495 (on file with staff).

[92] Demaneuf, G. (May 29, 2022). BSL Laboratories in Wuhan and their roles in coronaviruses research. Medium. https://gillesdemaneuf.medium.com/overview-of-biological-laboratories-in-wuhan-withtheir-roles-in-coronavirus-research-bca6c1cd1f74.

[93] *Id.*

[94] Qiu J. (June 1, 2020). How China's "Bat Woman" Hunted Down Viruses from SARS to the New Coronavirus. Scientific American. 322, 6, 24-32. doi:10.1038/scientificamerican0620-24.

[95] *Id.*, *see also*; Areddy JT. (Apr. 21, 2020). China Bat Expert Says Her Wuhan Lab Wasn't Source of New Coronavirus. Wall Street Journal. https://www.wsj.com/articles/chinas-bats-expert-says-her-wuhan-lab-wasnt-source-of-new-coronavirus-11587463204.

[96] *Id.*

[97] Editorial Board. We're still Missing the Origin Story of this Pandemic. China is Sitting on the Answers. The Post's View. Washington Post. https://www.washingtonpost.com/opinions/2021/02/05/coronavirus-origins-mystery-china/; *see also* Contributor, Anonymous & Bostickson, Billy & Demaneuf, Gilles. (2021). An Investigation into the WIV Databases that were Taken Offline. DOI:10.13140/RG.2.2.28029.08160

[98] *Supra*, note 92

[99] BurNIH-00000483-495 (on file with staff).

[100] *Supra*, note 18.

[101] Zhou, P., Yang, XL., Wang, XG. et al. (Feb. 3, 2020). A Pneumonia Outbreak Associated with a new Coronavirus of Probable Bat Origin. Nature. 579: 270–273. https://doi.org/10.1038/s41586-020-2012-7 https://www.nature.com/articles/s41586-020-2012-7.

[102] *Id.*

[103] Dou, Eva & Kuo, Lily. (Jun. 2, 2021). A Scientist Adventurer and China's "Bat Woman" are under Scrutiny as Coronavirus lab-leak Theory gets Another Look. Washington Post. https://www.washingtonpost.com/world/asia_pacific/coronavirus-bats-china-wuhan/2021/06/02/772ef984-beb2-11eb-922a-c40c9774bc48_story.html.

[104] *Id.*

[105] Woodward A. (Jun. 8, 2021). A 2019 Video Shows Scientists from the Wuhan CDC Collecting Samples in Bat caves — but the Agency hasn't Revealed any Findings. Business Insider. https://www.businessinsider.com/chinese-scientists-bat-caves-video-2021-6.

[106] *Supra*, note 91,

[107] Cohen J. Wuhan. (Jul. 31, 2020). Coronavirus Hunter Shi Zhengli Speaks out. Science. 369(6503):487-488. https://doi.org/10.1126/science.369.6503.487.

[108] *Supra*, note 91.

[109] *Id.*

[110] *Id.*

[111] *Id.*

[112] *Id.*

[113] *Id.*

[114] *Supra*, note 107.

[115] Li, N., Hu, L., Jin, A., & Li, J. (2019). Biosafety laboratory risk assessment. Journal of Biosafety and Biosecurity, 1(2), 90–92. https://doi.org/10.1016/j.jobb.2019.01.011.

[116] Wu, G. (2019). Laboratory biosafety in China: Past, present, and future. Biosafety and Health. https://doi.org/10.1016/j.bsheal.2019.10.003

[117] Zhou, P., Yang, X.-L., Wang, X.-G., Hu, B., Zhang, L., Zhang, W., Si, H.-R., Zhu, Y., Li, B., Huang, C.-L., Chen, H.-D., Chen, J., Luo, Y., Guo, H., Jiang, R.-D., Liu, M.-Q., Chen, Y., Shen, X.-R., Wang, X., & Zheng, X.-S. (2020). Addendum: A pneumonia outbreak associated with a new coronavirus of probable bat origin. Nature, 588(7836), E6–E6. https://doi.org/10.1038/s41586-020-2951-z

[118] *Id.*

[119] BurrNIH-0000016-54. (on file with staff).

[120] EcoHealth Alliance Project DEFUSE Proposal (on file with staff).

[121] *Id.*

[122] *Id.*

[123] *Id.*

[124] Li W, Wicht F, van Kuppeveld FJM, He Q, Rottier PJM, Bosch B-J. (May 13, 2015). A Single Point Mutation Creating a Furin Cleavage Site in the Spike Protein Renders Porcine Epidemic Diarrhea Coronavirus Trypsin Independent for cell entry and fusion. Journal of Virology. 2015 89(15) 80778081. https://doi.org/10.1128/jvi.00356-15.

[125] Sun, X., Belser, J. A., Yang, H., Pulit-Penaloza, J. A., Pappas, C., Brock, N., Zeng, H., Creager, H. M., Stevens, J., & Maines, T. R. (2019). Identification of key hemagglutinin residues responsible for cleavage, acid stability, and virulence of fifth-wave highly pathogenic avian influenza A(H7N9) viruses. Virology, 535, 232–240. https://doi.org/10.1016/j.virol.2019.07.012.

[126] Table of PRC Government Grants (on file with staff).

[127] *Supra*, note 107.

[128] *Supra*, note 91.

[129] Amrit, B.L.S. (Oct. 26, 2022). COVID-19: US NIH Partially Terminates Grant to EcoHealth Alliance. The Wire Science. https://science.thewire.in/the-sciences/us-national-institute-of-health-terminates-grant-to-nonprofit-that-worked-with-wuhan-institute/

[130] WIV patents on file with staff.

[131] Wuhan Institute of Virology (2019). Patent: Biological Safety Laboratory Exhaust System.(on file with staff).

[132] *Id.*

[133] *Id.*

[134] Wuhan Institute of Virology. Announcement of winning the bid for the procurement project of the environmental air disinfection system and the scalable automated sample storage management system of the Wuhan Institute of Virology, Chinese Academy of Sciences. (14 Aug. 2019). China Government Procurement Network. (on file with staff).

[135] *Id.*

[136] House Foreign Affairs Committee Report Minority Staff. (August 2021). The Origins of Covid-19: An Investigation of the Wuhan Institute of Virology. https://gop-foreignaffairs.house.gov/wp-content/uploads/2021/08/ORIGINS-OF-COVID-19-REPORT.pdf

[137] *Supra*, note 134.

[138] Henneman JR, McQuade EA, Sullivan RR, Downard J, Thackrah A, Hislop M. (Mar. 15, 2022).Analysis of Range and Use of a Hybrid Hydrogen Peroxide System for Biosafety 3 and Animal Biosafety Level 3 Agriculture laboratory Decontamination. Applied Biosafety. 27:1. https://doi.org/10.1089/apb.2021.0012. *See also*: Zhang S, Wu, J, Zhang E, et al. (Feb. 20, 2019). Research and Development of Airtight Biosafety Containment Facility for Stainless Steel Structures 2019. Journal of Biosafety and Biosecurity. 1: 56-62. https://doi.org/10.1016/j.jobb.2019.01.010. *See also*; Zhang H, Peng C, Liu B, Liu J, Zhiming Y, Shi, Z. (Mar. 1, 2018). Evaluation of MICRO-CHEM PLUS as a Disinfectant for Biosafety Level 4 Laboratory in China. Applied Biosafety Journal of ABSA International. 23(1): 32-38. http://doi.org/10.1177/153567601875*Id.*

[139] Jiali W. (September 16, 2019). Competitive Consultation on Central-air-Conditioning Renovation Project of Wuhan Institute of Virology, Chinese Academy Sciences China Government Procurement Network. (on file with staff).

[140] U.S. Department of Health and Human Services Centers for Disease Control and Prevention (CDC) Division of Select Agents and Toxins (DSAT). (2014). BSL-3/ABSL-3 HVAC and Facility Verification. https://www.cdc.gov/cpr/ipp/docs/policy_import_bsl3_absl3_verification.pdf.

[141] Wuhan Institute of Virology. (3 Dec. 2019). The Wuhan Institute of Virology of the Chinese Academy of Sciences Plans to use a Single-Source Procurement Method to Publicize the Procurement of Air Incineration Devices and Test Service Projects. China Government Procurement Network. (on file with staff).

[142] *Id.*

[143] Hanel E, Phillips GB, Gremillion GG. (1962). Technical Manuscript 1. Laboratory Design for Study of Infectious Disease. Office of the Safety Director US Army Chemical Corps Research and Development Command. Defense Technical Information Command Document #: 269-530 (on file with staff).

[144] Barbeito MS, Taylor LA, Seiders RW. (Mar. 16, 1968). Microbiological Evaluation of a Large-Volume Air Incinerator. Appl Microbiol. 16(3):490-495. https://doi.org/10.1128/am.16.3.490-495.1968

[145] Kuehne RW. (Sept. 26, 1973). Biological Containment Facility for Studying Infectious Disease. Appl Microbiol. 26(3):239-243. https://doi.org/10.1128/am.26.3.239-243.1973.

[146] Wuhan Institute of Virology. (Dec. 3 2019). The Wuhan Institute of Virology of the Chinese Academy of Sciences plans to use a single-source procurement method to publicize the procurement of air incineration devices and test service projects. https://archive.is/Jifqr#selection-229.0-229.197.

[147] Gao D, Zhang Q, Han K, Qian Q, Wenbo A. (December 11, 2019). Integrated Biological Sensor CN 201922213832.2. Google Patent. (on file with staff).

[148] Guo M, Yong M, Liu J, Huang X, Li X. (March 2019). Biosafety and Data Quality Considerations for Animal Experiments with Highly Infectious Agents at ABSL-3 Facilities. Journal of Biosafety and Biosecurity.1; 50-55. https://doi.org/10.1016/j.jobb.2018.12.011.

[149] *Id.*

[150] *Supra*, note 147.

[151] Jia W, Zhiming Y, Hao T, Jun L, Hao Q, Yi L, Lin W. Object surface disinfectant for high-grade biosafety laboratory and preparation method thereof. (on file with staff).

[152] *Id.*

[153] *Id.*

[154] Zhang *supra*, note 137.

[155] National Chemical Laboratories. (May 2022). Safety Data Sheet. https://www.nclonline.com/products/view/micro_chem_plus_#tab-safety.

[156] Email communication US Senate HELP Committee with Technical Representative National Chemical Laboratories May 11 2022 (on file with staff).

[157] Wuhan Institute of Virology. (09 July 2019). Communist Party Leaders Urge "Leapfrog Development" and Focus on "Stranglehold" Challenges: "Xiang Shuilun Examines the Wuhan Institute of Virology's Work of Establishing a 'Red Flag Party Branch'". *See also*: Wuhan Institute of Virology. (July 9, 2019). WIV Leaders Discuss and Correct "Shortcomings" & "Foundational Problems". *See also*: Wuhan Institute of Virology. (July 9,

2019). Wuhan Institute of Virology Organizes Centralized Study on the Educational Theme of 'Staying True to our Original Aspiration, Keeping Firmly in Mind our Mission'. (on file with staff).

[158] Yuan Zhiming. (Sept. 2019). Current Status and Future Challenges of High-Level Biosafety Laboratories in China. Journal of Biosafety and Biosecurity. 1:2. https://doi.org/10.1016/j.jobb.2019.09.005.

[159] Cao C. (2021 Jun 30). China's Evolving Biosafety/Biosecurity Legislations. J Law Biosci. 8(1):lsab020. doi: 10.1093/jlb/lsab020; *see also* Translate, C. L. (Oct. 18, 2020). Biosecurity Law of the P.R.C. China Law Translate. https://www.chinalawtranslate.com/en/biosecurity-law/

[160] Wuhan Institute of Virology. (June 11, 2019). Xiang Shuilun Examines the Wuhan Institute of Virology's Work of Establishing a 'Red Flag Party Branch'. (on file with staff).

[161] *Supra*, note 157.

[162] Wuhan Institute of Virology. (June 11, 2019). Xiang Shuilun Examines the Wuhan Institute of Virology's Work of Establishing a 'Red Flag Party Branch'. *See also*: Wuhan Institute of Virology. (July 9, 2019). Wuhan Institute of Virology Organizes Centralized Study on the Educational Theme of 'Staying True to our Original Aspiration, Keeping Firmly in Mind our Mission'. (on file with staff).

[163] *Id.*

[164] *Id.*

[165] Wuhan Institute of Virology. (July 30 2019). Wuhan Institute of Virology Convenes Study by the Party Committee's Plenary Central Group and Special Investigation and Study Meeting of the Educational Theme 'Never Forgetting our Original Aspiration and Keeping Firmly in Mind our Mission". (on file with staff).

[166] *Supra*, notes 97 & 136.

[167] *Id.*

[168] *Id.*

[169] *Id.*

[170] *Id.*

[171] Staff attempts to access the WIV database as recently as October 18, 2022 were unsuccessful. The website is http://batvirus.whiov.ac.cn/

[172] Wuhan Institute of Virology. (Nov. 12, 2019). Keep Firmly in Mind Your Responsibilities, Hold Fast to the Mission, Be a Pioneer for our Nation in the Realm of High-Level Biosafety – The Achievements of the Zhengdian Lab Party Branch of the Chinese Academy of Sciences Wuhan Institute of Virology. (on file with staff).

[173] *Id.*

[174] *Id.*

[175] *Id.*

[176] *Id.*

[177] *Id.*

[178] Wuhan Institute of Virology. (Nov. 21, 2019). Wuhan Institute of Virology Launches Training on Safety Work. (on file with staff).

[179] *Id.*

[180] National Human Genome Research Institute. (August 31, 2021,). COVID-19 mRNA Vaccine Production. Genome.gov. https://www.genome.gov/about-genomics/fact-sheets/COVID-19-mRNA-Vaccine-Production

[181] Campbell, C. (Aug. 24, 2020). Exclusive: Chinese Scientist Who First Sequenced COVID-19 Genome Speaks About Controversies Surrounding His Work. Time. https://time.com/5882918/zhang-yongzhen-interview-china-coronavirus-genome/.

[182] Krammer, F. (Sept. 23, 2020). SARS-CoV-2 Vaccines in Development. Nature. 586, 516–527. https://doi.org/10.1038/s41586-020-2798-3

[183] Cantrell, Jasper. (Mar. 16, 2020). How To: Recombinant Protein Construct Design Genetics And Genomics. Labroots. https://www.labroots.com/trending/genetics-and-genomics/17061/to-recombinant-protein-construct-design.

[184] United States Government Accountability Office. (2020). COVID-19 Federal Efforts Accelerate Vaccine and Therapeutic Development, but More Transparency Needed on Emergency Use Authorizations Report to Congressional Addressees. GAO-21-207. https://www.gao.gov/assets/gao-21-207.pdf.

[185] United States Government Accountability Office. (2021). OPERATION WARP SPEED Accelerated COVID- 19 Vaccine Development Status and Efforts to Address Manufacturing Challenges Report to Congressional Addressees. GAO-21-319. https://www.gao.gov/assets/gao-21-319.pdf.

[186] *Supra*, note 184.

[187] *Supra*, note 182.

[188] University of Oxford. (n.d.). About. Covid19vaccinetrial.co.uk. https://covid19vaccinetrial.co.uk/about.

[189] Bendix, S. N., Andrew Dunn, Aria. (Dec. 19, 2020). Moderna's Groundbreaking Coronavirus Vaccine was Designed in Just 2 Days. Business Insider. https://www.businessinsider.com/moderna-designed-coronavirus-vaccine-in-2-days-2020-11.

[190] *Supra*, note 185.

[191] University of Oxford. (January 20, 2022). A Phase I/II Study to Determine Efficacy, Safety and Immunogenicity of the Candidate Coronavirus Disease (COVID-19) Vaccine ChAdOx1 nCoV-19 in UK Healthy Adult Volunteers. Clinicaltrials.gov. https://clinicaltrials.gov/ct2/show/NCT04324606?term=phase+1%2C+ChAdOx1&cond=covid-19&draw=2&rank=1.

[192] Janssen Vaccines & Prevention B.V. (September 12, 2022). A Randomized, Double-blind, Placebo-controlled Phase 1/2a Study to Evaluate the Safety, Reactogenicity, and Immunogenicity of Ad26COVS1 in Adults Aged 18 to 55 Years Inclusive and Adults Aged 65 Years and Older. Clinicaltrials.gov. https://clinicaltrials.gov/ct2/show/study/NCT04436276.

[193] Zimmer C. (July 17, 2020). Inside Johnson & Johnson's Nonstop Hunt for a Coronavirus Vaccine. The New York Times. https://www.nytimes.com/2020/07/17/health/coronavirus-vaccine-johnson-janssen.html.

[194] *Supra*, note 185.

[195] Shulkin, D. (January 21, 2021). What Health Care Can Learn from Operation Warp Speed The "RAPID" Process Used in Operation Warp Speed Achieved Amazing Results at a Time of Great Need. NEJM Catalyst; New England Journal of Medicine. https://catalyst.nejm.org/doi/full/10.1056/CAT.21.0001.

[196] Slaoui, M., & Hepburn, M. (2020). Developing Safe and Effective Covid Vaccines — Operation Warp Speed's Strategy and Approach. New England Journal of Medicine. https://doi.org/10.1056/nejmp2027405.

[197] *Supra*, note 195.

[198] Le, Nhung. (Feb. 9, 2022). Meet the Scientist at the Center of the Covid lab leak Controversy. MIT Technology Review. https://www.technologyreview.com/2022/02/09/1044985/shi-zhengli-covid-lab-leak-wuhan/.

[199] Stevenson, A. (February 18, 2022). These Vaccines Have Been Embraced by the World. Why Not in China? The New York Times. https://www.nytimes.com/2022/02/18/business/china-coronavirus-vaccines.html

[200] An Y, Li S, Jin X, et al. (2022) A tandem-repeat dimeric RBD protein-based covid-19 vaccine zf2001 protects mice and nonhuman primates. Emerg Microbes Infect. 11(1):1058-1071. https://doi.org/10.1080/22221751.2022.2056524

[201] Ball P. (Dec. 18, 2020). The Lightning-Fast Quest for COVID Vaccines - and What it Means for Other Diseases. Nature. 589(7840):16-18. doi: 10.1038/d41586-020-03626-1. PMID: 33340018.

[202] Zhou Y, Zhao G, Gu H; Sun S, He L, Li Y, Han G, Lang X, Liu J, Geng S, Sheng X. (Feb. 24, 2020). Preparation of COVID-19 Vaccine Comprising RBD Domain-Fc Fusion Protein for Prevention and Therapy of SARS-CoV-2 Virus Infection. State Intellectual Property Office of the People's Republic of China. CN111333704A. https://qxb-img-osscache.qixin.com/patents_pdf_new/b916b8b4b29175cb63ab43dabe6ae785.pdf.

[203] *Id.*

[204] Pan, X., Zhou, P., Fan, T., Wu, Y., Zhang, J., Shi, X., Shang, W., Fang, L., Jiang, X., Shi, J., Sun, Y., Zhao, S., Gong, R., Chen, Z., & Xiao, G. (2020). Immunoglobulin Fragment F(ab')$_2$ Against RBD Potently Neutralizes SARS-CoV-2 in Vitro. Antiviral Research. 182 104868. https://doi.org/10.1016/j.antiviral.2020.104868.

[205] Sun S, He L, Zhao Z, et al. (Mar. 21, 2021). Recombinant Vaccine Containing an RBD-Fc Fusion Induced Protection against SARS-CoV-2 in Nonhuman Primates and Mice. Cell Mol Immunol. 18(4):1070-1073. https://doi.org/10.1038/s41423-021-00658-z.

[206] Halperin SA, Ye L, MacKinnon-Cameron D, Smith B, Cahn PE, Ruiz-Palacios GM, Ikram A, Lanas F, Lourdes Guerrero M, Muñoz Navarro SR, Sued O, Lioznov DA, Dzutseva V, Parveen G, Zhu F, Leppan L, Langley JM, Barreto L, Gou J, Zhu T. CanSino COVID-19 Global Efficacy Study Group. (Dec. 23, 2021). Final efficacy Analysis, Interim Safety Analysis, and Immunogenicity of a Single Dose of Recombinant Novel Coronavirus Vaccine (adenovirus type 5 vector) in Adults 18 years and Older: an International, Multicentre, Randomised, Double-blinded, Placebo-Controlled Phase 3 trial. Lancet. 2022399(10321):237-248.

[207] *Supra*, note 43.

[208] Rambaut, A., Holmes, E. C., O'Toole, Á., Hill, V., McCrone, J. T., Ruis, C., du Plessis, L., & Pybus, O. G. (Jul. 15, 2020). A dynamic nomenclature proposal for SARS-CoV-2 lineages to assist genomic epidemiology. *Nature microbiology*, 5(11), 1403–1407. https://doi.org/10.1038/s41564-020-0770-5.

[209] *Supra*, note 43.

[210] *Supra*, note 94.

[211] BurNIH-00000483-495 (on file with staff).

[212] *Supra*, note 97.

[213] Cohen J. (Jul. 31, 2020). Wuhan Coronavirus Hunter Shi Zhengli speaks out. Science. 369(6503), 487–488. https://doi.org/10.1126/science.369.6503.487.

[214] *Id.*

[215] *Id.*

[216] *Supra*, note 158.

[217] Wuhan Institute of Virology. (June 21, 2019). Wuhan Institute of Virology Convenes Promotion Meeting for Work on the Educational Theme of 'Staying True to our Original Aspiration, Keeping Firmly in Mind our Mission' and a Study Session of the Expanded Party Committee Central Group. (on file with staff).

[218] Wuhan Institute of Virology. (July 30, 2019). Wuhan Institute of Virology Convenes Study by the Party Committee's Plenary Central Group and Special Investigation and Study Meeting of the Educational Theme 'Never Forgetting our Original Aspiration and Keeping Firmly in Mind our Mission'. (on file with staff).

[219] *Supra*, note 159.

[220] *Supra*, note 172.

[221] Wuhan Institute of Virology. (Nov. 21, 2019). Wuhan Institute of Virology Launches Training on Safety Work. (on file with staff).

[222] *Supra*, note 17.

[223] *Supra*, note 86.

[224] *Supra*, note 43.

[225] *Supra*, note 2; *see also* Sachs, J. D., Karim, S. S. A., Aknin, L., Allen, J., Brosbøl, K., Colombo, F., Barron, G. C., Espinosa, M. F., Gaspar, V., Gaviria, A., Haines, A., Hotez, P. J., Koundouri, P., Bascuñán, F. L., Lee, J.-K., Pate, M. A., Ramos, G., Reddy, K. S., Serageldin, I., & Thwaites, J. (2022). The Lancet Commission on lessons for the future from the COVID-19 pandemic. The Lancet, 0(0). https://doi.org/10.1016/S0140-6736(22)01585-9. *See also*: Office of the Director of National Intelligence. (2021). Updated Assessment on COVID-19 Origins. https://www.dni.gov/files/ODNI/documents/assessments/Declassified-Assessment-on-COVID-19-Origins.pdf.

www.ingramcontent.com/pod-product-compliance
Lightning Source LLC
Chambersburg PA
CBHW081403270326

41930CB00015B/3401